London Mathematical Society Lecture Note Series. 5

J.T.KNIGHT

Commutative Algebra

CAMBRIDGE AT THE UNIVERSITY PRESS 1971

CAMBRIDGE UNIVERSITY PRESS
Cambridge, New York, Melbourne, Madrid, Cape Town, Singapore, São Paulo

Cambridge University Press
The Edinburgh Building, Cambridge CB2 8RU, UK

Published in the United States of America by Cambridge University Press, New York

www.cambridge.org
Information on this title: www.cambridge.org/9780521081931

First published 1971
Re-issued in this digitally printed version 2007

A catalogue record for this publication is available from the British Library

Library of Congress Catalogue Card Number: 76-152625

ISBN 978-0-521-08193-1 paperback

These notes were put together with scissors and paste from my manuscript, with as many errors corrected as my audience and I could find. I should like to thank everyone who helped, and especially Barry Tennison.

James T. Knight

Dr J. T. Knight died in April 1970 from injuries received in a motor accident. In preparing the notes for publication, I have made some slight amendments; I hope that the original flavour has been preserved. My thanks are due to the London Mathematical Society, and particularly to Professors J. W. S. Cassels and P. M. Cohn for their help.

Barry R. Tennison

October 1970

J. T. Knight

James Thomson Knight was born in Glasgow in 1942. It was in his home city that he began his university education, after he had gained first place in the Glasgow University Entrance Bursary Competition in 1960. Four years later he graduated with first class honours in Mathematics, and, for achieving first place in the final examination, was awarded the Bryce Fellowship.

He then went to Trinity College, Cambridge, where he worked in number theory and commutative algebra under the supervision of Dr. (now Professor) J. W. S. Cassels, leading in 1967 to the degree of Ph. D., and election to a Junior Research Fellowship at Churchill College, Cambridge.

After spending the next year as Lecturer in Mathematics at the United College, Chinese University of Hong Kong, he returned to take up the Fellowship in Cambridge. His research interests were in commutative algebra and ring theory. The list below summarises his published work, which, apart from the first item, some work in physical chemistry done with ICI while he was an undergraduate at Glasgow University, has all appeared in the Proceedings of the Cambridge Philosophical Society.

Dr. Knight died in a car accident in April 1970 while visiting a friend in Ireland.

J. T. Knight's Publications

A kinetic theory of droplet coalescence with application to emulsion stability (with R. A. W. Hill), Trans. Faraday Soc. 61 (1965) 170-181.

Quadratic forms over R(t), Proc. Camb. Phil. Soc. 62 (1966) 197-205.

Binary integral quadratic forms over R(t), Proc. Camb. Phil. Soc. 62 (1966) 433-440.

Riemann surfaces of field extensions, Proc. Camb. Phil. Soc. 65 (1969) 635-650.

Some rings of interest in the study of places, Proc. Camb. Phil. Soc. 68 (1970) 255-264.

On epimorphisms of non-commutative rings, Proc. Camb. Phil. Soc. 68 (1970) 589-600.

A note on residually finite groups, Proc. Camb. Phil. Soc. 69 (1971).

B. R. T.

Contents

1. Preliminaries

1.1 INTRODUCTION

The subject matter of commutative algebra is the common ground of geometry and arithmetic. As these subjects have grown more abstract, more and more common ground has been found, and commutative algebra has grown very large.

In chapters 1 to 5 we shall cover some general material; in chapter 6 there will be some results useful in geometry; and in chapters 7 and 8 some arithmetical results.

An undergraduate algebra course should be all you need to follow this course.

1.2 DEFINITIONS AND RECAPITULATIONS

By 'ring' we mean 'commutative ring with a one'; for example, the *zero ring* $\{0\}$:

+	0
0	0

.	0
0	0

By 'ring morphism' we mean a one-preserving homomorphism.

Let A and B be rings. We say that A is a *subring* of B iff $A \subseteq B$ and the inclusion map: $A \to B$ is a ring morphism. For example $\{0\}$ is *not* a subring of the integers $\mathbf{Z}$; but $\mathbf{Z}$ *is* a subring of the rationals $\mathbf{Q}$.

We say that a ring A is <u>integral</u> iff

(i) A is non-zero

(ii) for $x, y \in A$, if $xy = 0$, then $x = 0$ or $y = 0$.

We say that a ring A is a <u>field</u> iff

(i) A is non-zero

(ii) for all $x \in A$ with $x \neq 0$ there exists $y \in A$ such that $xy = 1$.

(We write x^{-1} for this unique y.)

Let us write A^* for the set of all non-zero elements of a ring A. Then A is integral (resp. a field) iff A^* is a semigroup (resp. a group) under multiplication. Thus every field is integral.

For a non-zero ring A we define $\mathfrak{U}(A) = \{x \in A; xy = 1$ for some $y \in A\}$; $\mathfrak{U}(A)$ is a group under multiplication and is called the <u>group of units</u> of A. For a field $\mathfrak{U}(A) = A^*$.

Let $\mathfrak{a}$ be an ideal of A. The following statements are equivalent:

(i) $\mathfrak{a} \subset A$

(ii) $1 \notin \mathfrak{a}$

(iii) $\mathfrak{a} \cap \mathfrak{U}(A) = \emptyset$

and we call such an ideal <u>proper</u>. Note that for $x \in A$ we have:

$x \notin \mathfrak{U}(A)$ iff xA is proper.

A subset S is of a ring A is called <u>multiplicative</u> iff

(i) $1 \in S$

(ii) $x, y \in S$ implies $xy \in S$;

for example: $\{1\}$; $\mathfrak{U}(A)$; A itself.

Let $\mathfrak{p}$ be an ideal of A. The following conditions are equivalent:

(i) $A \setminus \mathfrak{p}$ is multiplicative

(ii) $A/\mathfrak{p}$ is integral.

We call such an ideal prime. Note that it must be proper. For example, A is integral iff $\{0\}$ is a prime ideal.

Let $\mathfrak{m}$ be an ideal of A. The following conditions are equivalent:

(i) $\mathfrak{m}$ is a maximal element of the set of all proper ideals ordered by inclusion

(ii) $A/\mathfrak{m}$ is a field.

We call such an ideal maximal. Every maximal ideal is prime.

1.2.1 (Proposition). Let A be a ring; $S \subseteq A$ be multiplicative; and $\mathfrak{a}$ be an ideal of A with $\mathfrak{a} \cap S = \emptyset$. Then there is an ideal $\mathfrak{p}$ of A, maximal among those ideals $\mathfrak{b}$ with $\mathfrak{b} \supseteq \mathfrak{a}$ and $\mathfrak{b} \cap S = \emptyset$; and any such $\mathfrak{p}$ is prime.

Proof. Let

$$X = \{\mathfrak{b} \text{ an A-ideal: } \mathfrak{b} \supseteq \mathfrak{a} \text{ and } \mathfrak{b} \cap S = \emptyset\}.$$

Then $\mathfrak{a} \in X \neq \emptyset$. Let $Y \subseteq X$ be non-empty and totally ordered by inclusion. Then $\cup(Y) \in X$ and by Zorn's lemma X has maximal elements. If $\mathfrak{p}$ is such an element, $1 \notin \mathfrak{p}$; and if $x, y \notin \mathfrak{p}$ and $xy \in \mathfrak{p}$, then $\mathfrak{p} + xA$, $\mathfrak{p} + yA \supset \mathfrak{p}$ so that $\mathfrak{p} + xA$, $\mathfrak{p} + yA \notin X$. Thus there exist $s, t \in S$; $p, q \in \mathfrak{p}$;

$a, b \in A$ such that $s = p + xa$ and $t = q + yb$. Then $st = pq + xaq + ybp + abxy \in \mathfrak{p} \cap S$, a contradiction. Thus $A \setminus \mathfrak{p}$ is multiplicative and $\mathfrak{p}$ is prime. □

1.2.1.1 (Corollary). A non-zero ring has a maximal ideal.

Proof. Take $S = \{1\}$ and $\mathfrak{a} = \{0\}$. □

1.2.1.2 (Corollary). Let A be a non-zero ring. Then

$$\mathfrak{u}(A) = A \setminus \bigcup_{\mathfrak{m}} \mathfrak{m}$$

where the union is taken over all maximal ideals $\mathfrak{m}$ of A.

Proof. One way is immediate. Conversely let $x \in A \setminus \mathfrak{u}(A)$. Then $xA \subset A$ and there is a maximal ideal $\mathfrak{m}$ of A with $xA \subseteq \mathfrak{m}$: so that $x \in \mathfrak{m}$. □

We say that $x \in A$ is nilpotent iff $x^n = 0$ for some $n \in \omega$. We write $\mathfrak{n}(A)$ for the ideal of all nilpotent elements of A, and call $\mathfrak{n}(A)$ the nilradical of A.

1.2.1.3 (Corollary). Let A be a non-zero ring. Then

$$\mathfrak{n}(A) = \bigcap_{\mathfrak{p}} \mathfrak{p}$$

taken over all prime ideals $\mathfrak{p}$ of A.

Proof. If $x^n = 0$, then $x^n \in \mathfrak{p}$ so $x \in \mathfrak{p}$ for all prime ideals $\mathfrak{p}$. Conversely if $x^n \neq 0$ for all $n \in \omega$, then $S = \{x^n : n \in \omega\}$ is multiplicative and $S \cap \{0\} = \emptyset$; thus $S \cap \mathfrak{p} = \emptyset$ for some prime ideal $\mathfrak{p}$ of A, and in particular $x \notin \mathfrak{p}$. □

It may be that a non-zero ring A has only one maximal ideal $\mathfrak{m}(A)$. In this case we call A local; $\mathfrak{m}(A)$ is its greatest proper ideal; and $\mathfrak{u}(A) = A \setminus \mathfrak{m}(A)$. We call $A/\mathfrak{m}(A)$ the residual field $\kappa(A)$ of A. For example, a field is local and is its own residual field.

If $\mathfrak{a}$ and $\mathfrak{b}$ are ideals of a ring A we define the ideal $\mathfrak{a}\mathfrak{b}$ as follows:

$$\mathfrak{a}\mathfrak{b} = \{ \sum_{i=1}^{m} a_i b_i : a_i \in \mathfrak{a} \text{ and } b_i \in \mathfrak{b} \}.$$

If $\mathfrak{p}$ is prime and $\mathfrak{p} \supseteq \mathfrak{a}\mathfrak{b}$, then plainly $\mathfrak{p} \supseteq \mathfrak{a}$ or $\mathfrak{p} \supseteq \mathfrak{b}$. Thus $\supseteq$ is like 'divides'.

If $(\mathfrak{a}_\lambda)_{\lambda \in \Lambda}$ is a family of ideals of A, we define $\sum_{\lambda \in \Lambda} \mathfrak{a}_\lambda$ to be the ideal of all $\sum_{\lambda \in \Lambda} a_\lambda$ for families $(a_\lambda)_{\lambda \in \Lambda}$ in $\prod_{\lambda \in \Lambda} \mathfrak{a}_\lambda$ with $a_\lambda = 0$ for all but a finite number of $\lambda \in \Lambda$.

We call the set spec(A) of all prime ideals of a ring A the spectrum of A. It is non-empty iff A is non-zero. For ideals $\mathfrak{a}$ of A we define

$V(\mathfrak{a}) = \{ \mathfrak{p} \in \operatorname{spec}(A) : \mathfrak{p} \supseteq \mathfrak{a} \}$ so that

$$\operatorname{spec}(A) = V(\{0\})$$

$$\emptyset = V(A)$$

$$V(\mathfrak{a}) \cup V(\mathfrak{b}) = V(\mathfrak{a}\mathfrak{b})$$

$$\bigcap_{\lambda \in \Lambda} V(\mathfrak{a}_\lambda) = V(\sum_{\lambda \in \Lambda} \mathfrak{a}_\lambda)$$

and the $V(\mathfrak{a})$ are thus the closed sets for a topology on spec(A), called the <u>Zariski topology</u>. It is rarely Hausdorff as we shall see.

Exercise. spec(A) is compact.

If $f: A \to B$ is a ring morphism we define $F: \mathrm{spec}(B) \to \mathrm{spec}(A)$ by $F(\mathfrak{q}) = f^{-1}[\mathfrak{q}]$. Then F is continuous because $F^{-1}[V(\mathfrak{a})] = V(Bf[\mathfrak{a}])$ for any ideal $\mathfrak{a}$ of A. Thus spec is a contravariant functor from the category of rings to the category of topological spaces (see Appendix 1). If f is onto, the homomorphism theorems for rings show that F is an embedding.

1.3 MODULES

Let A be a ring. An A-<u>module</u> M is an additive group (M, +) together with a multiplication: $A \times M \to M$ such that

$$\lambda(m + m') = \lambda m + \lambda m'$$

$$(\lambda + \lambda')m = \lambda m + \lambda' m$$

$$(\lambda\lambda')m = \lambda(\lambda' m)$$

$$1m = m$$

for all $\lambda, \lambda' \in A$ and $m, m' \in M$. For example, if A is a field, an A-module is just a vector space over A.

Most definitions and some theorems for modules are just like those for vector spaces. For example, if $(M_\lambda)_{\lambda \in \Lambda}$ is a family of A-modules, the direct sum $\bigoplus_{\lambda \in \Lambda} M_\lambda$ consists of all families $(m_\lambda)_{\lambda \in \Lambda} \in \prod_{\lambda \in \Lambda} M_\lambda$ of finite support (that is with $m_\lambda = 0$ for all but a finite number of $\lambda \in \Lambda$), and with componentwise addition and multiplication.

A ring is a module over itself; and its submodules are its ideals.

Let L and M be A-modules and N be an additive group (resp. an A-module). Let $g: L \times M \to N$. We say that g is bilinear iff

(i) $g(l + l', m) = g(l, m) + g(l', m)$

(ii) $g(l, m + m') = g(l, m) + g(l, m')$

(iii) $g(\lambda l, m) = g(l, \lambda m)$.

(resp. $g(\lambda l, m) = g(l, \lambda m) = \lambda g(l, m)$)

for $l, l' \in L$; $m, m' \in M$; and $\lambda \in A$.

On the set $\mathbf{Z}^{(L \times M)}$ of all functions $f: L \times M \to \mathbf{Z}$ with $f(l, m) = 0$ for all but a finite number of $(l, m) \in L \times M$, define addition thus:

$$(f + f')(l, m) = f(l, m) + f'(l, m)$$

Thus $\mathbf{Z}^{(L \times M)}$ becomes an additive group. Let W be the subgroup generated by

$$f_{(l + l', m)} - f_{(l, m)} - f_{(l', m)}$$

$$f_{(l, m + m')} - f_{(l, m)} - f_{(l, m')}$$

$$f_{(\lambda l, m)} - f_{(l, \lambda m)}$$

for all $l, l' \in L$; $m, m' \in M$; and $\lambda \in A$:
where f_α is the function such that

$$f_\alpha(\beta) = 1 \text{ if } \beta = \alpha$$
$$= 0 \text{ if } \beta \neq \alpha \ .$$

We call $\mathbf{Z}^{(L \times M)}/W$ the <u>tensor product</u> $L \otimes_A M$ of L and M, and write $l \otimes m$ for $f_{(l, m)} + W$. Thus if $x \in L \otimes_A M$ then $x = \sum_{i=1}^{r} l_i \otimes m_i$ for $l_i \in L$ and $m_i \in M$; and $\otimes : L \times M \to L \otimes_A M$ is bilinear.

$L \otimes_A M$ has the following universal property: if N is an additive group and $g: L \times M \to N$ is bilinear, then there is one and only one group morphism: $L \otimes_A M \to N$ such that the diagram

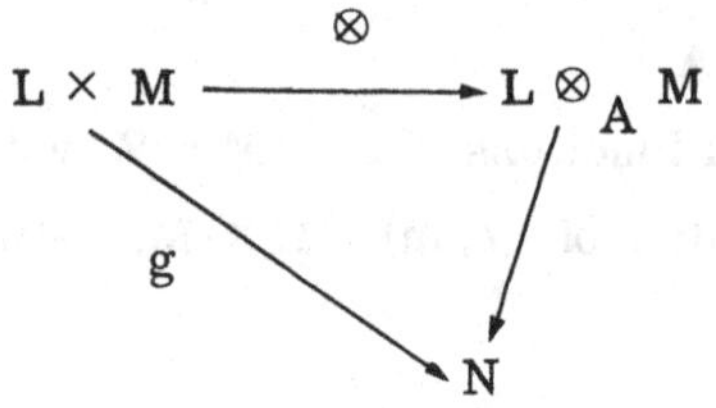

commutes: namely the function:

$$f + W \mapsto \sum_{\substack{l \in L \\ m \in M}} f(l, m)\, g(l, m) \ .$$

We use this universal property first of all to make $L \otimes_A M$ into an A-module: multiplication by $\lambda \in A$ is the unique group morphism: $L \otimes_A M \to L \otimes_A M$ such that $l \otimes m \mapsto (\lambda l) \otimes m$. With this structure, if N is an A-module and $g: L \times M \to N$ is bilinear, the unique group morphism:

$L \otimes_A M \to N$ such that

$$\begin{array}{ccc} L \times M & \to & L \otimes_A M \\ & \searrow & \downarrow \\ & & N \end{array}$$

commutes is also a module morphism (that is, a linear map).

Next if $g:L \to L'$ and $h:M \to M'$ are A-module morphisms, we define $g \otimes h$ to be the unique group morphism $L \otimes_A M \to L' \otimes_A M'$ such that $l \otimes m \mapsto g(l) \otimes h(m)$. Of course $g \otimes h$ is also an A-module morphism.

If $(M_\lambda)_{\lambda \in \Lambda}$ is a family of A-modules, there is a natural module isomorphism:

$$L \otimes_A (\bigoplus_{\lambda \in \Lambda} M_\lambda) \to \bigoplus_{\lambda \in \Lambda} (L \otimes_A M_\lambda) \quad \text{given by}$$

$l \otimes (m_\lambda)_{\lambda \in \Lambda} \mapsto (l \otimes m_\lambda)_{\lambda \in \Lambda}$ (its inverse comes from putting together the natural maps: $L \otimes_A M_\mu \to L \otimes_A (\bigoplus_{\lambda \in \Lambda} M_\lambda)$ for each $\mu \in \Lambda$).

Let M be an A-module and $(m_\lambda)_{\lambda \in \Lambda}$ be a family in M. We say that $(m_\lambda)_{\lambda \in \Lambda}$ generates (resp. bases) M iff for all $m \in M$ there is a family (resp. a unique family) $(\xi_\lambda)_{\lambda \in \Lambda}$ of finite support and such that $m = \sum_{\lambda \in \Lambda} \xi_\lambda m_\lambda$.

If M has a base we say that M is free: for example vector spaces are free.

If some finite family generates M, we say that M is of finite type.

If $f:A \to B$ is a ring morphism we make B into an A-module by defining $\lambda b = f(\lambda)b$ for $\lambda \in A$ and $b \in B$. If also M is an A-module we make $B \otimes_A M$ into a B-module by defining $\mu(b \otimes m) = (\mu b) \otimes m$ for $\mu, b \in B$ and $m \in M$. (We call $B \otimes_A M$ the B-ification of M.)

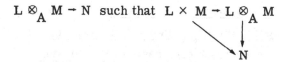

1.3.1 (Lemma). Let M be an A-module and $f: A \to B$ be a ring morphism. Suppose $(m_\lambda)_{\lambda \in \Lambda}$ generates (resp. bases) M. Then $(1 \otimes m_\lambda)_{\lambda \in \Lambda}$ generates (resp. bases) the B-module $B \otimes_A M$.

Proof. The generates part is immediate.

Suppose therefore that $(m_\lambda)_{\lambda \in \Lambda}$ bases M. For $\mu \in \Lambda$ define the A-module morphism $p_\mu: M \to A$ by

$$p_\mu\left(\sum_{\lambda \in \Lambda} \xi_\lambda m_\lambda\right) = \xi_\mu$$

and the B-module morphism $h_\mu: B \otimes_A M \to B$ by $h_\mu(b \otimes m) = bf(p_\mu(m))$. Then

$$h_\mu\left(\sum_{\lambda \in \Lambda} \eta_\lambda (1 \otimes m_\lambda)\right) = \sum_{\lambda \in \Lambda} \eta_\lambda\, f(p_\mu(m_\lambda))$$

$$= \eta_\mu$$

for any family $(\eta_\lambda)_{\lambda \in \Lambda}$ in B of finite support. Thus if $x \in B \otimes_A M$, there is a unique family $(h_\lambda(x))_{\lambda \in \Lambda}$ in B of finite support such that

$$x = \sum_{\lambda \in \Lambda} h_\lambda(x)\,(1 \otimes m_\lambda)\ . \square$$

1.3.1.1 (Corollary). Let A be a non-zero ring and M be a free A-module of finite type. Then there exists $n \in \omega$ such that if $(m_\lambda)_{\lambda \in \Lambda}$ bases M, then Λ has n elements.

Proof. A has a maximal ideal $\mathfrak{m}$ and $A/\mathfrak{m} \otimes_A M$ is a vector space of finite type. Let n be the dimension of $A/\mathfrak{m} \otimes_A M$. Since $(1 \otimes m_\lambda)_{\lambda \in \Lambda}$ bases $A/\mathfrak{m} \otimes_A M$, Λ has n elements.□

We call n the rank of M.

Strange to say 1.3.1.1 breaks down for non-commutative rings.

There is a proof of 1.3.1.1 which does not use the axiom of choice.

Exercise. Let $(m_\lambda)_{\lambda \in \Lambda}$ (resp. $(n_\mu)_{\mu \in M}$) generate (resp. base) the A-module X (resp. Y). Show that $(m_\lambda \otimes n_\mu)_{(\lambda, \mu) \in \Lambda \times M}$ generates (resp. bases) the A-module $X \otimes_A Y$.

An A-module M is called Noetherian iff it satisfies the three equivalent conditions:

(i) every submodule of M is of finite type

(ii) every non-empty set of sub-modules of M has a maximal element

(iii) every increasing sequence $N_1 \subseteq N_2 \subseteq \dots$ of sub-modules of M is eventually constant.

For example, a vector space of finite type is Noetherian.

If a ring A is a Noetherian A-module it is called a Noetherian ring.

1.3.2 (Proposition). Let M be an A-module and N be a submodule. Then M is Noetherian iff N and M/N are Noetherian.

Proof. Suppose N and M/N are Noetherian and let $L_1 \subseteq L_2 \subseteq \ldots$ be an increasing sequence of submodules of M. Then there exists $i \in \omega$ such that $L_i + N = L_j + N$ and $L_i \cap N = L_j \cap N$ for all $j > i$. From the commutative diagram

$$\begin{array}{ccc} N/L_i \cap N & \xrightarrow{\text{iso}} & (L_i + N)/L_i \\ \downarrow 1 & & \downarrow \text{inclusion} \\ N/L_j \cap N & \xrightarrow{\text{iso}} & (L_j + N)/L_j \end{array}$$

we see that $L_i = L_j$.

The converse is immediate. □

1.3.2.1 (Corollary). *If* $M = \sum_{i=1}^{n} M_i$ *is a finite sum of Noetherian submodules* M_i, *then* M *is Noetherian.* □

1.3.2.2 (Corollary). *If* M *is an* A-*module and* A *is Noetherian,* M *is Noetherian iff it is of finite type.* □

1.3.2.3 (Corollary). *A submodule of an* A-*module of finite type over a Noetherian ring* A *is itself of finite type.* □

Exercise. If M and N are Noetherian A-modules, so is $M \otimes_A N$.

A sequence

$$M_0 \xrightarrow{f_0} M_1 \xrightarrow{f_1} \ldots \xrightarrow{f_{n-1}} M_n$$

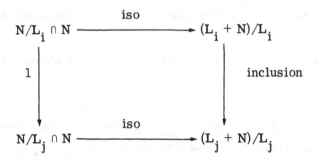

of A-module morphisms is called exact iff

$$f_i[M_i] = \mathrm{Ker}(f_{i+1}) \text{ for } i = 0, \ldots, n-2.$$

For example $0 \to M \to N$ is exact iff $M \to N$ is injective; $M \to N \to 0$ is exact iff $M \to N$ is onto; and $0 \to M \xrightarrow{f} N \xrightarrow{g} L \to 0$ is exact iff f is injective and $N/f[M] \xrightarrow{g} L$ is a well-defined isomorphism.

Exercise. A functor T: A-modules $\to$ additive groups preserves all exact sequences iff it preserves all exact sequences of three modules iff it preserves all exact sequences of the form $0 \to M \to N \to L \to 0$. (Break up $M \xrightarrow{f} N \xrightarrow{g} L$ as follows:

$$0 \to \ker f \to M \to f[M] \to 0$$

$$0 \to f[M] \to N \to N/f[M] \to 0$$

$$0 \to N/f[M] \to L \to L/g[N] \to 0$$

and apply T.) We call such a functor T exact.

An A-algebra M is an A-module M together with a bilinear multiplication: $M \times M \to M$ which makes M into a ring. Thus the map: $A \to M$ given by $\lambda \mapsto \lambda 1$ is a ring morphism. Conversely any ring morphism: $A \to B$ makes B into an A-algebra.

If M and N are A-algebras there is one and only one bilinear multiplication on $M \otimes_A N$ such that

$$(m \otimes n)(m' \otimes n') = (mm') \otimes (nn')$$

for $m, m' \in M$ and $n, n' \in N$. We call $M \otimes_A N$ with this algebra structure the Kronecker product of M and N. (In fact it is the sum of M and N in the category of A-algebras.)

1.3.3 (Theorem). (I. S. Cohen) Let A be a ring. Then A is Noetherian iff every prime ideal is of finite type.

Proof. If A is Noetherian, every ideal is of finite type. Conversely, suppose A is not Noetherian; then the set X of ideals not of finite type is non-empty; and if $Y \subseteq X$ is totally ordered by $\subseteq$, $\cup Y \in X$. By Zorn, X has a maximal element $\mathfrak{p}$. We show $\mathfrak{p}$ prime.

First, $1 \notin \mathfrak{p}$. Suppose $\mathfrak{p}$ is not prime, and let $a, b \in A\setminus \mathfrak{p}$ be such that $ab \in \mathfrak{p}$. Then $\mathfrak{p} + aA \supset \mathfrak{p}$ so $\mathfrak{p} + aA \notin X$. Hence $\mathfrak{p} + aA = x_1A + \ldots + x_rA$ for some $x_i = p_i + ay_i \in \mathfrak{p} + aA$. Now consider

$$\mathfrak{b} = \{y \in A: ay \in \mathfrak{p}\}.$$

Then $\mathfrak{p} \subseteq \mathfrak{b}$ and $b \in \mathfrak{b}$, so $\mathfrak{b} \supseteq \mathfrak{p} + bA \supset \mathfrak{p}$. Hence $\mathfrak{b} \notin X$, and so

$$\mathfrak{b} = z_1A + \ldots + z_sA \text{ for some } z_1, \ldots, z_s \in \mathfrak{b}.$$

Clearly

$$\mathfrak{p} \supseteq p_1A + \ldots + p_rA + az_1A + \ldots + az_sA$$

and if $p \in \mathfrak{p}$ then

$$p = \sum_{i=1}^{r} a_i(p_i + ay_i) \qquad (*)$$

for some $a_i \in A$. Hence

$$a\left(\sum_{i=1}^{r} a_i y_i\right) = p - \sum_{i=1}^{r} a_i p_i \in \mathfrak{p}$$

and so $\sum_{i=1}^{r} a_i y_i \in \mathfrak{b}$. Therefore $\sum_{i=1}^{r} a_i y_i = \sum_{j=1}^{s} b_j z_j$, say.

Hence from (*)

$$p = \sum_{i=1}^{r} a_i p_i + \sum_{j=1}^{s} b_j (az_j)$$

$$\in p_1 A + \dots + p_r A + az_1 A + \dots + az_s A .$$

Thus $\mathfrak{p} = p_1 A + \dots + p_r A + az_1 A + \dots + az_s A$, but $\mathfrak{p} \in X$, a contradiction. □

2. Flatness

2.1 PROJECTIVE MODULES

If $f:M \to N$ is a morphism of A-modules and P is an A-module, we can define an A-module morphism: $\mathrm{Hom}_A(P, M) \to \mathrm{Hom}_A(P, N)$ by $\phi \mapsto f \circ \phi$; and $M \mapsto \mathrm{Hom}_A(P, M)$ becomes a covariant functor, written $\mathrm{Hom}_A(P, \ )$.

2.1.1 (Proposition). (i) *Let* $0 \to L \xrightarrow{f} M \xrightarrow{g} N$ *be an exact sequence of A-modules. Then for any A-module* P *the sequence* $0 \to \mathrm{Hom}_A(P, L) \to \mathrm{Hom}_A(P, M) \to \mathrm{Hom}_A(P, N)$ *is exact.*

(ii) *The following conditions on an A-module* P *are equivalent:*

(a) $\mathrm{Hom}_A(P, \ )$ *is exact,*

(b) *given a diagram*

$$\begin{array}{ccccc} & & P & & \\ & & \downarrow & & \\ M & \to & N & \to & 0 \end{array}$$

of A-modules where the row is exact, there is a morphism: $P \to M$ *such that*

$$\begin{array}{ccc} & & P \\ & \swarrow & \downarrow \\ M & \to & N \end{array}$$

commutes,

(c) *there is an A-module* Q *such that* $P \oplus Q$ *is free.*

Proof. (i) If $\phi:P \to L$ is such that $f \circ \phi = 0$, then $\phi = 0$ because f is injective.

If $\psi:P \to M$ is such that $g \circ \psi = 0$, then for all $x \in P$ there is a unique $\phi(x) \in L$ such that $f(\phi(x)) = \psi(x)$: thus $\phi \in \mathrm{Hom}_A(P, L)$ and $\phi \mapsto \psi \in \mathrm{Hom}_A(P, M)$. Conversely if

$\phi: P \to L$, then $\phi \mapsto g \circ f \circ \psi = 0 \in \mathrm{Hom}_A(P, N)$.

(ii) (b) says that we can put 0 on the ends of the sequence in (i). Thus (see last exercise) (a) and (b) are equivalent.

Suppose (b) holds. Let F be a free module (for example $A^{(P)}$) and $f: F \to P$ be onto. Then there is a morphism $g: P \to F$ such that

$$\begin{array}{ccc} & & P \\ & \swarrow^{g} & \downarrow 1_P \\ F & \xrightarrow{f} & P \end{array}$$

commutes. Let us map $F \to P \oplus \mathrm{Ker}\, f$ by $x \mapsto (f(x), x - g(f(x)))$; and $P \oplus \mathrm{Ker}\, f \to F$ by $(x, y) \mapsto g(x) + y$; these morphisms are inverse to each other and thus $P \oplus \mathrm{Ker}\, f$ is isomorphic to F.

Conversely suppose (c) holds and let $(x_\lambda)_{\lambda \in \Lambda}$ base $P \oplus Q$ and let $P \xrightarrow{i} P \oplus Q \xrightarrow{p} P$ be the natural maps. Consider

$$\begin{array}{ccccc} & & P & & \\ & & \downarrow g & & \\ M & \xrightarrow{f} & N & \to & 0 \end{array}$$

; for $\lambda \in \Lambda$ choose $m_\lambda \in M$ such that $f(m_\lambda) = g(p(x_\lambda))$ and define $h: P \oplus Q \to M$ by

$$h\Big(\sum_{\lambda \in \Lambda} \xi_\lambda x_\lambda\Big) = \sum_{\lambda \in \Lambda} \xi_\lambda m_\lambda .$$

Then

$$\begin{array}{ccc} & & P \\ & \swarrow^{h \circ i} & \downarrow g \\ M & \xrightarrow{f} & N \end{array}$$

commutes. □

We call an A-module P which satisfies (a), (b) and (c) <u>projective</u>. For example, all modules over fields are free and therefore projective; $\mathbf{Z}/2\mathbf{Z}$ is <u>not</u> a projective $\mathbf{Z}$-module.

An easy modification of 2. 1. 1 shows that P is projective and of finite type iff there is a module Q such that $P \oplus Q$ is free and of finite type. Note that if M is free and $f: P \to M$ and

$g:M \to P$ are such that $g \circ f = 1_P$, then P is projective.

2.2 FLAT MODULES

(These are rather like projective modules, only $\otimes_A$ is used instead of Hom_A.)

Let $f:M \to N$ be a morphism of A-modules and E be an A-module. Thus $1_E \otimes f: E \otimes_A M \to E \otimes_A N$, and $M \mapsto E \otimes_A M$ becomes a covariant functor, which we write $E \otimes_A$.

2.2.1 (Proposition). <u>Let $L \xrightarrow{f} M \xrightarrow{g} N \to 0$ be an exact sequence of A-modules and E be an A-module. Then $E \otimes_A L \to E \otimes_A M \to E \otimes_A N \to 0$ is exact.</u>

Proof. Let $x \in E \otimes_A N$. Then $x = \sum_{i=1}^{r} e_i \otimes n_i$ for some $e_i \in E$ and $n_i \in N$. Moreover $n_i = g(m_i)$ for some $m_i \in M$; thus $x = (1_E \otimes g)(\sum_{i=1}^{r} e_i \otimes m_i)$.

Plainly $(1_E \otimes g) \circ (1_E \otimes f) = 1_E \otimes (g \circ f) = 0$. Thus we may define:

$$E \otimes_A M \Big/ \mathrm{Im}(1_E \otimes f) \xrightarrow{1_E \otimes g} E \otimes_A N$$

and it remains to show that this is an isomorphism. To do this we construct its inverse: we map $e \otimes n \mapsto e \otimes m + \mathrm{Im}(1_E \otimes f)$ for any $m \in M$ with $g(m) = n$. This is in order, for if $g(m) = g(m')$ $= n$, then $m - m' = f(l)$ for some $l \in L$ and:

$$e \otimes m - e \otimes m' = e \otimes f(l) \in \mathrm{Im}(1_E \otimes f) . \square$$

2.2.1.1 (Corollary). *Let* E *be an* A*-module. The following conditions are equivalent:*

(i) $E \otimes_A$ *is exact,*

(ii) *if* M *is an* A*-module and* N *is a submodule of finite type, then* $E \otimes_A N \to E \otimes_A M$ *is injective.*

Proof. Suppose (ii) holds and let $f: L \to M$ be an injective morphism of A-modules. Let $x \in E \otimes_A L$ be such that $x \mapsto 0 \in E \otimes_A M$. Then $x = \sum_{i=1}^{r} e_i \otimes l_i$ for $e_i \in E$ and $l_i \in L$. Let N be the submodule of M generated by $(f(l_i))_{1 \le i \le r}$. We have

$$\begin{array}{ccc} & \overset{f}{\nearrow} & M \\ L & & \uparrow \\ & \underset{f^{-1}}{\nwarrow} & N \end{array}$$

commuting; thus

$$\begin{array}{ccc} & \nearrow & E \otimes_A M \\ E \otimes_A L & & \uparrow \text{inj} \\ & \nwarrow & E \otimes_A N \end{array}$$

commutes; so that $\sum_{i=1}^{r} e_i \otimes f(l_i) \mapsto 0 \in E \otimes_A M$ and thus $\sum_{i=1}^{r} e_i \otimes f(l_i) = 0$ and:

$$x = (1_E \otimes f^{-1}) \left(\sum_{i=1}^{r} e_i \otimes f(l_i) \right) = 0 .$$

Thus $E \otimes_A L \to E \otimes_A M$ is injective and $E \otimes_A$ is exact.

The converse is immediate. □

If E satisfies the conditions of 2.2.1.1 we say that E is *flat*.

Exercises. If $(E_\lambda)_{\lambda \in \Lambda}$ is a family of flat modules, $\bigoplus_{\lambda \in \Lambda} E_\lambda$ is flat.

If E and F are flat, so is $E \otimes_A F$.

If $A \to B \to C$ are ring morphisms, B is a flat A-module, and C is a flat B-module, then C is a flat A-module.

Examples. Since $A \otimes_A M \xrightarrow{\text{iso}} M$ by $\lambda \otimes m \mapsto \lambda m$, A is a flat A-module. Thus (see exercise) every free module is flat.

If P is a projective A-module, then $P \oplus Q$ is free for some module Q; so if $M \to N$ is an injective morphism of A-modules we have the commutative diagram:

$$\begin{array}{ccc}
(P \oplus Q) \otimes_A M & \xrightarrow{\text{inj}} & (P \oplus Q) \otimes_A N \\
\downarrow \text{iso} & & \downarrow \text{iso} \\
(P \otimes_A M) \oplus (Q \otimes_A M) & \longrightarrow & (P \otimes_A N) \oplus (Q \otimes_A N) \\
\uparrow \text{inj} & & \uparrow \text{inj} \\
P \otimes_A M & \longrightarrow & P \otimes_A N
\end{array}$$

Thus $P \otimes_A M \to P \otimes_A N$ is injective and P is flat.

We say that an A-module E is <u>flat for</u> an A-module M iff for every submodule N of M of finite type we have:

$$E \otimes_A N \xrightarrow{\text{inj}} E \otimes_A M .$$

Thus E is flat iff E is flat for every M.

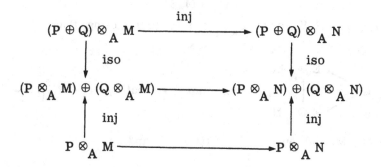

2.2.2 (Lemma). (i) *If* E *is flat for* M *and* N *is a submodule of* M, *then* E *is flat for* M/N.

(ii) *If* E *is flat for* M_λ *for each* $\lambda \in \Lambda$, *then* E *is flat for* $\bigoplus_{\lambda \in \Lambda} M_\lambda$.

Proof. (i) The argument of 2.2.1.1 shows that for every submodule L of M we have $E \otimes_A L \xrightarrow{\text{inj}} E \otimes_A M$. Thus if L/N is a submodule of M/N (of finite type), we have the commutative diagram

$$\begin{array}{ccccccc} N & \longrightarrow & L & \longrightarrow & L/N & \longrightarrow & 0 \\ \big\downarrow{\scriptstyle 1_N} & & \big\downarrow & & \big\downarrow & & \\ N & \longrightarrow & M & \longrightarrow & M/N & \longrightarrow & 0 \end{array}$$

with exact rows. Therefore by 2.2.1

$$\begin{array}{ccccccc} E \otimes_A N & \longrightarrow & E \otimes_A L & \longrightarrow & E \otimes_A (L/N) & \longrightarrow & 0 \\ \big\downarrow{\scriptstyle 1_{E \otimes_A N}} & & \big\downarrow{\scriptstyle \text{inj}} & & \big\downarrow & & \\ E \otimes_A N & \longrightarrow & E \otimes_A M & \longrightarrow & E \otimes_A (M/N) & \longrightarrow & 0 \end{array}$$

commutes and has exact rows. Let $x \in E \otimes_A (L/N)$ and $x \mapsto 0 \in E \otimes_A (M/N)$. There exists $y \in E \otimes_A L$ such that $y \mapsto x$. Suppose $y \mapsto z \in E \otimes_A M$. Then $z \mapsto 0 \in E \otimes_A (M/N)$ so $w \mapsto z$ for some $w \in E \otimes_A N$. Let $w \mapsto u \in E \otimes_A L$, then $u, y \mapsto z \in E \otimes_A M$ so $u = y$ and $w \mapsto x$ along the top. Thus $x = 0$: that is, E is flat for M/N.

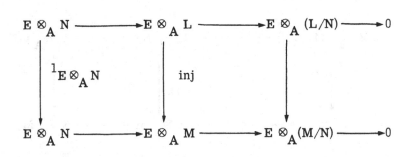

(ii) First let $\Lambda = \{1, 2\}$ and $M = M_1 \oplus M_2$. We have the exact sequences

$$0 \to M_1 \xrightarrow{i} M \xrightarrow{p} M_2 \to 0$$

$$0 \to E \otimes_A M_1 \to E \otimes_A M \to E \otimes_A M_2 \to 0$$

since $E \otimes_A M$ is isomorphic to $(E \otimes_A M_1) \oplus (E \otimes_A M_2)$.

Let N be a submodule of M (of finite type): we have the commutative diagram

$$\begin{array}{ccccccccc} 0 & \longrightarrow & M_1 \cap N & \xrightarrow{i} & N & \xrightarrow{p} & p[N] & \longrightarrow & 0 \\ & & \downarrow & & \downarrow & & \downarrow & & \\ 0 & \longrightarrow & M_1 & \xrightarrow{i} & M & \xrightarrow{p} & M_2 & \longrightarrow & 0 \end{array}$$

with exact rows: thus the diagram

$$\begin{array}{ccccc} E \otimes_A (M_1 \cap N) & \longrightarrow & E \otimes_A N & \longrightarrow & E \otimes_A p[N] \\ \downarrow \text{inj} & & \downarrow & & \downarrow \text{inj} \\ E \otimes_A M_1 & \xrightarrow{\text{inj}} & E \otimes_A M & \longrightarrow & E \otimes_A M_2 \end{array}$$

commutes and has exact rows. Let $x \in E \otimes_A N$ and $x \mapsto 0 \in E \otimes_A M$. Thus $x \mapsto 0 \in E \otimes_A p[N]$ so $y \mapsto x$ for some $y \in E \otimes_A (M_1 \cap N)$. But $y \mapsto 0 \in E \otimes_A M$; thus $y = 0$ and $x = 0$. Therefore E is flat for M.

Next by induction we extend the result to any finite Λ.

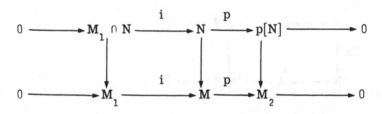

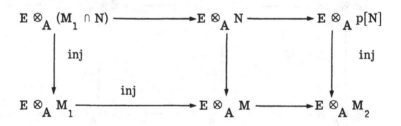

Finally let N be a submodule of finite type of $M = \bigoplus_{\lambda \in \Lambda} M_\lambda$. Then $N \subseteq \bigoplus_{\lambda \in \Lambda_1} M_\lambda$ for some *finite* $\Lambda_1 \subseteq \Lambda$. Let $M_1 = \bigoplus_{\lambda \in \Lambda_1} M_\lambda$ and $M_2 = \bigoplus_{\lambda \in \Lambda \setminus \Lambda_1} M_\lambda$ so that $M = M_1 \oplus M_2$. As before $E \otimes_A M_1 \xrightarrow{\text{inj}} E \otimes_A M$; and we have shown that $E \otimes_A N \xrightarrow{\text{inj}} E \otimes_A M_1$; so $E \otimes_A N \xrightarrow{\text{inj}} E \otimes_A M$ and E is flat for M. □

2.2.3 (Theorem). *E is a flat A-module iff* $E \otimes_A \mathfrak{a} \xrightarrow{\text{inj}} E$ *by* $x \otimes \lambda \mapsto \lambda x$ *for any ideal* $\mathfrak{a}$ *of* A *of finite type.*

Proof. We must show that if E is flat for A (the assertion on the right) then E is flat for any A-module M. Firstly E is flat for $\bigoplus_{m \in M} A$ by 2.2.2; and $(\xi_m)_{m \in M} \mapsto \sum_{m \in M} \xi_m m$ maps $\bigoplus_{m \in M} A$ *onto* M; thus E is flat for M by 2.2.2. □

2.2.3.1 (Corollary). *Let A be a Bezout ring (i.e. an integral ring in which every ideal of finite type is principal) and E be an A-module. Then E is flat iff E is torsion-free (i.e.* $\lambda m = 0$ *for* $\lambda \in A$ *and* $m \in E$ *implies* $\lambda = 0$ *or* $m = 0$).

Proof. Suppose E is torsion-free. If $\lambda \in A$ we have the commutative diagram

$$\begin{array}{ccc} E \otimes_A A & \xrightarrow{\;x \otimes \mu \;\mapsto\; \mu x\;} & E \\ \text{onto} \Big\downarrow \times \lambda & & \text{inj} \Big\downarrow \times \lambda \\ E \otimes_A \lambda A & \longrightarrow & E \end{array}$$

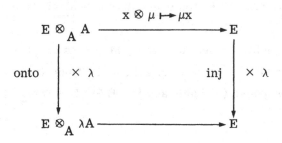

If $x \in E \otimes_A \lambda A$ and $x \mapsto 0 \in E$, there exists $y \in E \otimes_A A$ such that $y \mapsto x$; and $y \mapsto 0 \in E$; so $y = 0$ and $x = 0$. Thus $E \otimes_A \lambda A \xrightarrow{\text{inj}} E$. If $\lambda = 0$ there is nothing to prove.

Conversely let $\lambda \in A^*$. Then $A \xrightarrow{\text{inj}} A$ by $\mu \mapsto \mu\lambda$; so $E \otimes_A A \xrightarrow[\times \lambda]{\text{inj}} E \otimes_A A$. But $E \otimes_A A$ is isomorphic to E; so $E \xrightarrow[\times \lambda]{\text{inj}} E$: that is, E is torsion-free. □

Examples. If A is a field, every A-module is flat. (Choice is not used here.)

A $\mathbf{Z}$-module is flat iff it is torsion-free; thus $\mathbf{Q}$ is a flat $\mathbf{Z}$-module but $\mathbf{Z}/7\mathbf{Z}$ is not.

2.3 FAITHFULLY FLAT MODULES

2.3.1 (Theorem). *Let E be an A-module. The following three statements are equivalent:*

(i) $L \xrightarrow{f} M \xrightarrow{g} N$ *is an exact sequence of A-modules iff* $E \otimes_A L \to E \otimes_A M \to E \otimes_A N$ *is an exact sequence of A-modules;*

(ii) E *is flat; and for any A-module* M, *if* $E \otimes_A M = 0$, *then* $M = 0$;

(iii) E *is flat and* $\mathfrak{m} E \subset E$ *for every maximal ideal* $\mathfrak{m}$ *of* A.

Proof. Suppose (i) holds. Then $0 \to E \otimes_A M \to 0$ is exact if $E \otimes_A M = 0$, so $0 \to M \to 0$ is exact and $M = 0$. Thus (ii) holds.

Conversely suppose (ii) holds and let $E \otimes_A L \to E \otimes_A M \to$

$E \otimes_A N$ be exact. We have

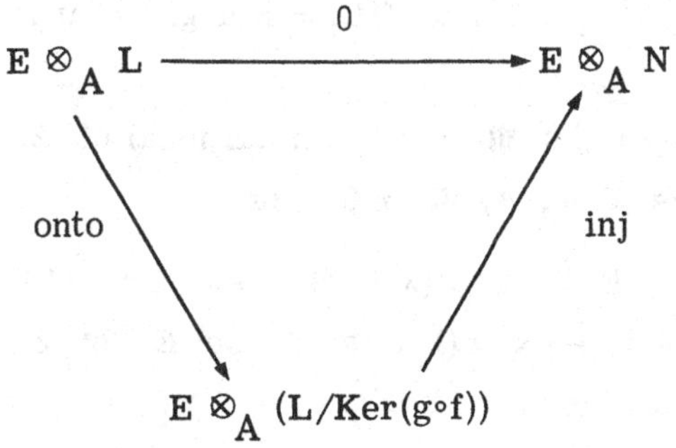

commuting; thus $E \otimes_A (L/\mathrm{Ker}(g \circ f)) = 0$ and $g \circ f = 0$.

Let $X = \mathrm{Im}(E \otimes_A L \to E \otimes_A M)$

$= \mathrm{Ker}(E \otimes_A M \to E \otimes_A N)$

and consider the commutative diagram:

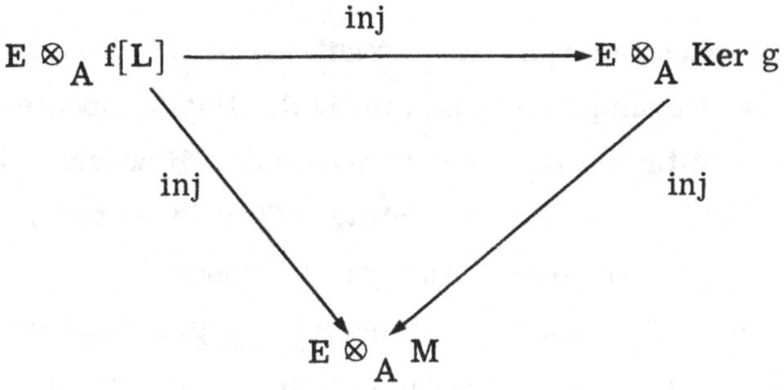

In $E \otimes_A M$ we have:

$$X \subseteq \mathrm{Im}(E \otimes_A f[L]) \subseteq \mathrm{Im}(E \otimes_A \mathrm{Ker}\, g) \subseteq X$$

so that $E \otimes_A f[L] \xrightarrow{\text{iso}} E \otimes_A \mathrm{Ker}\, g$. But

$$0 \to E \otimes_A f[L] \to E \otimes_A \operatorname{Ker} g \to E \otimes_A \operatorname{Ker} g/f[L] \to 0$$

is exact: so $E \otimes_A \operatorname{Ker} g/f[L] = 0$ and $f[L] = \operatorname{Ker} g$. That is, (i) holds.

Suppose (ii) holds and let $\mathfrak{m}$ be a maximal ideal of A. Then $A/\mathfrak{m} \neq 0$ and thus $E \otimes_A A/\mathfrak{m} \neq 0$. But $E \otimes_A A/\mathfrak{m} \xrightarrow{\text{iso}} E/\mathfrak{m}E$ by $x \otimes (\lambda + \mathfrak{m}) \mapsto \lambda x + \mathfrak{m}E$ (whose inverse is $x + \mathfrak{m}E \mapsto x \otimes (1 + \mathfrak{m})$); so $E/\mathfrak{m}E \neq 0$: that is, $\mathfrak{m}E \subset E$ and (iii) holds.

Conversely let (iii) hold and let $M \neq 0$: say $m \in M$ with $m \neq 0$. Let $\mathfrak{a}$ be the kernel of the map: $A \to M$ by $\lambda \mapsto \lambda m$ and let $\mathfrak{m}$ be a maximal ideal of A with $\mathfrak{m} \supseteq \mathfrak{a}$. Then $\mathfrak{a}E \subseteq \mathfrak{m}E \subset E$ so that $E \otimes_A A/\mathfrak{a} \xrightarrow{\text{iso}} E/\mathfrak{a}E \neq 0$. However $A/\mathfrak{a} \xrightarrow{\text{inj}} M$; so that $E \otimes_A M \neq 0$ since E is flat. Thus (ii) holds. □

If E satisfies the equivalent conditions of 2.3.1 we call E <u>faithfully flat.</u> For example, A is a faithfully flat A-module; and a vector space is faithfully flat iff it is non-zero. However, $\mathbf{Q}$ is <u>not</u> a faithfully flat $\mathbf{Z}$-module since $(p\mathbf{Z})\mathbf{Q} = \mathbf{Q}$ if p is prime.

Let M be an A-module. An exact sequence $L_1 \to L_0 \to M \to 0$ of A-modules is called a <u>finite presentation</u> of M iff L_0 and L_1 are free and of finite type. For example, if M is projective and of finite type, $M \oplus L$ is free and of finite type for some L, so $N \to L \to 0$ is exact for some free N of finite type and $N \to M \oplus L \to M \to 0$ is a finite presentation of M. Again, if A is Noetherian and M is of finite type, $L \to M \to 0$ is exact for some free L of finite type: and if $N = \operatorname{Ker}(L \to M)$, N is of finite type by 1.3.2.3, so that $Q \to N \to 0$ is exact for some free Q of finite type and $Q \to L \to M \to 0$ is a finite presentation of M.

2.3.2 (Lemma). *Let* $0 \to M \xrightarrow{f} N \xrightarrow{g} P \to 0$ *be an exact sequence of A-modules. Suppose that* N *is of finite type and* P *is of finite presentation. Then* M *is of finite type.*

Proof. Let $L_1 \xrightarrow{\phi} L_0 \xrightarrow{\psi} P \to 0$ be a finite presentation of P. We have the commutative diagram with exact rows and columns:

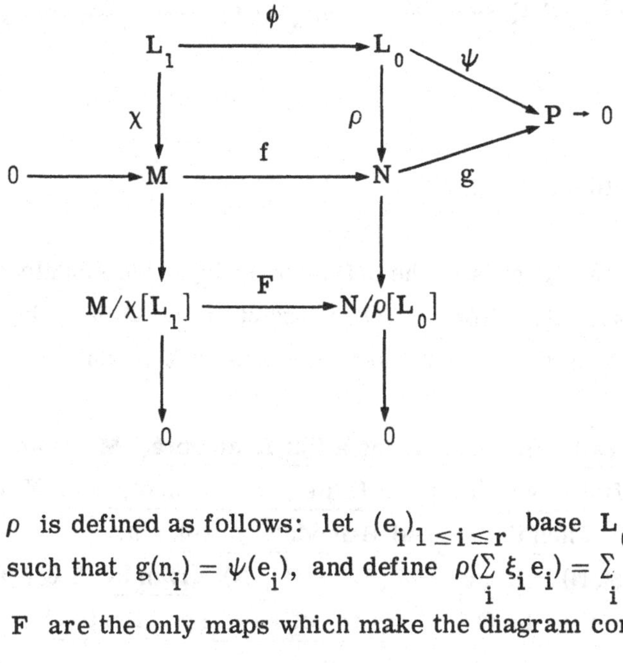

where ρ is defined as follows: let $(e_i)_{1 \leq i \leq r}$ base L_0, choose $n_i \in N$ such that $g(n_i) = \psi(e_i)$, and define $\rho(\sum_i \xi_i e_i) = \sum_i \xi_i n_i$; and χ and F are the only maps which make the diagram commute, namely

$$\chi = \{(x, y) : x \in L_1;\ y \in M;\ \rho(\phi(x)) = f(y)\}$$

$$F = \{(x + \chi[L_1],\ f(x) + \rho[L_0]) : x \in M\}.$$

Then F is an isomorphism, for $F^{-1}: N/\rho[L_0] \to M/\chi[L_1]$ consists of all pairs $(y + \rho[L_0],\ z + \chi[L_1])$ such that $y \in N$ and $z \in M$

and $g(y) = \psi(u)$ and $f(z) = y - \rho(u)$ for some $u \in L_0$. Thus $M/\chi[L_1]$ is of finite type; and L_1 is of finite type; so M is of finite type. □

2.3.3 (Lemma). Let $L \xrightarrow{f} M \xrightarrow{g} N \to 0$ be an exact sequence of A-modules and Q be an A-module. Then

$$0 \to \mathrm{Hom}_A(N, Q) \xrightarrow{\phi \mapsto \phi \circ g} \mathrm{Hom}_A(M, Q) \xrightarrow{\psi \mapsto \psi \circ f} \mathrm{Hom}_A(L, Q)$$

is exact.

Proof. Just like 2.1.1(i). □

This lemma leads to the definition of injective modules as modules Q such that $\mathrm{Hom}_A(\ , Q)$ is exact, e.g. $\mathbf{Q}/\mathbf{Z}$ is injective but not flat; $\mathbf{Z}$ is flat but not injective (each as $\mathbf{Z}$-modules).

2.3.4 (Proposition). Let B be a flat A-algebra; M be an A-module of finite type (resp. of finite presentation); and N be an A-module. Then the natural B-module morphism: $B \otimes_A \mathrm{Hom}_A(M, N) \to \mathrm{Hom}_B(B \otimes_A M, B \otimes_A N)$ given by $b \otimes f \mapsto \phi$, where $\phi: c \otimes m \mapsto bc \otimes f(m)$, is injective (resp. bijective).

Proof. Let T (resp. T') be the functor $B \otimes_A \mathrm{Hom}_A(\ , N)$ (resp. $\mathrm{Hom}_B(P \otimes_A\ , B \otimes_A N)$). Note that if $(M_i)_{i \in I}$ is a finite family of A-modules, there are natural B-module morphisms:

$$T(\bigoplus_{i \in I} M_i) \to \bigoplus_{i \in I} T(M_i) \quad \text{and} \quad T'(\bigoplus_{i \in I} M_i) \to \bigoplus_{i \in I} T'(M_i)$$

such that

$$\begin{array}{ccc} T(\bigoplus_{i \in I} M_i) & \longrightarrow & T'(\bigoplus_{i \in I} M_i) \\ \downarrow & & \downarrow \\ \bigoplus_{i \in I} T(M_i) & \longrightarrow & \bigoplus_{i \in I} T'(M_i) \end{array}$$

commutes.

Suppose first that $M = A$. The diagram

$$\begin{array}{ccc} T(A) & \longrightarrow & T'(A) \\ \downarrow{\scriptstyle \text{iso}} & & \downarrow{\scriptstyle \text{iso}} \\ B \otimes_A N & \xrightarrow{1_B \otimes_A N} & B \otimes_A N \end{array}$$

commutes, where $T(A) \to B \otimes_A N$ is given by $b \otimes f \mapsto b \otimes f(1)$ (and its inverse by $b \otimes n \mapsto b \otimes \phi$ where $\phi: a \mapsto an$) and $T'(A) \to B \otimes_A N$ is given by $f \mapsto f(1 \otimes 1)$ (and its inverse by $b \otimes n \mapsto f$ where $f(c \otimes a) = bc \otimes an$). Thus $T(A) \to T'(A)$ is bijective. Therefore (taking account of the first paragraph) $T(M) \to T'(M)$ is bijective if M is free and of finite type.

(Note: we have not yet used the flatness of B.)

Now let $L_1 \to L_0 \to M \to 0$ be exact and L_0 (resp. L_1 and L_0) be free and of finite type. By 2. 3. 3 and since B is flat we have the commutative diagram

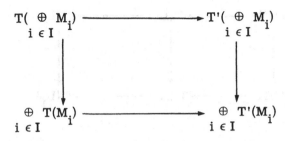

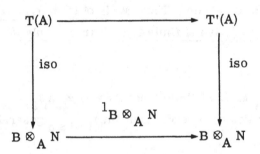

$$\begin{array}{ccccccc}
0 & \longrightarrow & T(M) & \longrightarrow & T(L_0) & \longrightarrow & T(L_1) \\
 & & \downarrow & & \downarrow u & & \downarrow v \\
0 & \longrightarrow & T'(M) & \longrightarrow & T'(L_0) & \longrightarrow & T'(L_1)
\end{array}$$

with exact rows; and u (resp. u and v) is (resp. are) bijective. It is immediate that $T(M) \to T'(M)$ is injective (resp. bijective). □

2.3.5 (Proposition). Let B be a faithfully flat (resp. flat) A-algebra and M be an A-module. Then M is of finite type (resp. of finite presentation; projective of finite type) if (resp. only if) $B \otimes_A M$ is so too.

Proof. Suppose $B \otimes_A M$ is of finite type; then $(1 \otimes m_i)_{1 \le i \le n}$ generates $B \otimes_A M$ for some $m_i \in M$. Let $(m_i)_{1 \le i \le n}$ generate the submodule N of M; then

$$B \otimes_A N \xrightarrow{\text{onto}} B \otimes_A M \text{ so that } M = N .$$

Suppose $B \otimes_A M$ is of finite presentation; then M is of finite type so $L \to M \to 0$ is exact for some free L of finite type. Let $N = \mathrm{Ker}(L \to M)$ so that $0 \to B \otimes_A N \to B \otimes_A L \to B \otimes_A M \to 0$ is exact. By 2.3.2 $B \otimes_A N$ is of finite type so that N is of finite type and $R \to N \to 0$ is exact for some free R of finite type.. Thus $R \to L \to M \to 0$ is a finite presentation of M.

Suppose $B \otimes_A M$ is projective and of finite type. Then M is of finite presentation. Let $L \xrightarrow{\text{onto}} N$ be a morphism of A-modules and consider the commutative diagram

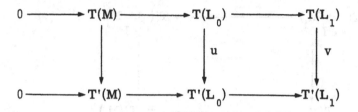

$$\begin{array}{ccc}
B \otimes_A \mathrm{Hom}_A(M, L) & \xrightarrow{\text{iso}} & \mathrm{Hom}_B(B \otimes_A M,\ B \otimes_A L) \\
\downarrow & & \downarrow \\
B \otimes_A \mathrm{Hom}_A(M, N) & \xrightarrow{\text{iso}} & \mathrm{Hom}_B(B \otimes_A M,\ B \otimes_A N)
\end{array}$$

in which the horizontal isomorphisms are given in 2.3.4. The vertical arrow on the right is onto because $B \otimes_A L \xrightarrow{\text{onto}} B \otimes_A N$ and $B \otimes_A M$ is projective; thus the vertical arrow on the left is onto. But B is faithfully flat: so $\mathrm{Hom}_A(M, L) \xrightarrow{\text{onto}} \mathrm{Hom}_A(M, N)$ and M is projective (and of finite type).

The converse is immediate. □

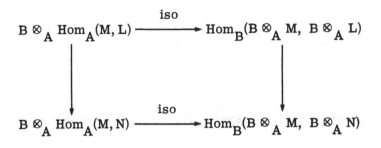

3. Fractions

3.1 RINGS, MODULES AND ALGEBRAS OF FRACTIONS

Let A be a ring and $S \subseteq A$ be multiplicative. On $S \times A$ define an equivalence $\sim$ as follows:

$$(s, a) \sim (s', a') \text{ iff } t(s'a - sa') = 0 \text{ for some } t \in S.$$

We write $S^{-1}A$ for $S \times A/\sim$ and $\frac{a}{s}$ for the equivalence class of (s, a). We make $S^{-1}A$ into an A-algebra by defining

$$\frac{a}{s} + \frac{a'}{s'} = \frac{s'a + sa'}{ss'}$$

$$\frac{a}{s}\frac{a'}{s'} = \frac{aa'}{ss'}$$

$$\lambda \frac{a}{s} = \frac{\lambda a}{s} \quad \text{for } \lambda \in A .$$

The natural map: $A \to S^{-1}A$ is given by $a \mapsto \frac{a}{1}$.

The A-algebra $S^{-1}A$ has the following universal property: if $f: A \to B$ is a ring morphism such that $f[S] \subseteq \mathfrak{u}(B)$, there is one and only one ring morphism: $S^{-1}A \to B$ such that

$$\begin{array}{ccc} A & \to & S^{-1}A \\ & {\scriptstyle f}\searrow & \swarrow \\ & & B \end{array}$$

commutes, namely $\frac{a}{s} \mapsto f(s)^{-1} f(a)$. (In other words, there is a unique A-algebra morphism: $S^{-1}A \to B$.)

The one of $S^{-1}A$ is $\frac{1}{1}$ and its zero is $\frac{0}{1}$. Thus $\frac{a}{s} = 0$ iff $ta = 0$ for some $t \in S$; and $S^{-1}A = 0$ iff $0 \in S$.

The map: $A \to S^{-1}A$ is an isomorphism iff $S \subseteq \mathfrak{u}(A)$.

Let $T = \{t \in A:\ a \in A^* \Longrightarrow ta \in A^*\}$, the multiplicative subset of non-divisors of zero in A. We call $T^{-1}A = \text{tot}(A)$ the total ring of fractions of A.

The kernel of the map: $A \to S^{-1}A$ is the ideal $\{a \in A:\ ta = 0 \text{ for some } t \in S\}$; thus $A \overset{\text{inj}}{\to} S^{-1}A$ iff $S \subseteq T$. In particular $A \overset{\text{inj}}{\to} \text{tot}(A)$.

If A is integral $T = A^*$ and tot(A) is the field of fractions of A. Conversely if tot(A) is integral it is a field and A is integral.

Exercise. If A is Noetherian, so is $S^{-1}A$.

3.1.1 (Proposition). Let A be a ring and $S \subseteq A$ be multiplicative. Then $\text{spec}(S^{-1}A) \to \text{spec}(A)$ is an increasing embedding whose image is

$$X = \{\mathfrak{p} \in \text{spec}(A):\ \mathfrak{p} \cap S = \emptyset\}$$

and whose inverse is also increasing.

Proof. The map: $\text{spec}(S^{-1}A) \to X \subseteq \text{spec}(A)$ is given by $\mathfrak{q} \mapsto \{p \in A: \frac{p}{1} \in \mathfrak{q}\}$; its inverse: $X \to \text{spec}(S^{-1}A)$ is given by $\mathfrak{p} \mapsto \{\frac{p}{s}:\ p \in \mathfrak{p} \text{ and } s \in S\}$.

The rest is immediate. □

Let $f: A \to B$ be a ring morphism and $S \subseteq A$ and $T \subseteq B$ be multiplicative and such that $f[S] \subseteq T$. Then there is one and only one ring morphism:

$S^{-1}A \to T^{-1}B$ such that

$$\begin{array}{ccc} A & \longrightarrow & B \\ \big\downarrow & & \big\downarrow \\ S^{-1}A & \longrightarrow & T^{-1}B \end{array}$$

commutes: namely $\frac{a}{s} \mapsto \frac{f(a)}{f(s)}$.

For example let A be integral and $k = \mathrm{tot}(A)$ be its field of fractions, so that we may identify A with the subring $\{\frac{a}{1}: a \in A\}$ of k. If we take $f = 1_A$ and $T = A^*$ we can also identify $S^{-1}A$ with the subring $\{s^{-1}a: a \in A \text{ and } s \in S\}$ of k, for any multiplicative $S \subseteq A$ with $0 \notin S$.

Let M be an A-module and $S \subseteq A$ be multiplicative. We define the A-module $S^{-1}M$ analogously:

$$\frac{m}{s} = \frac{m'}{s'} \text{ iff } t(s'm - sm') = 0 \text{ for some } t \in S$$

$$\frac{m}{s} + \frac{m'}{s'} = \frac{s'm + sm'}{ss'}$$

$$\lambda \frac{m}{s} = \frac{\lambda m}{s} \quad \text{for } \lambda \in A .$$

We can also make $S^{-1}M$ into an $S^{-1}A$ module by defining

$$\frac{a}{s}\,\frac{m}{t} = \frac{am}{st} \quad .$$

3.1.2 (Lemma). *The map* $f: S^{-1}A \otimes_A M \to S^{-1}M$ *given by* $\frac{a}{s} \otimes m \mapsto \frac{am}{s}$ *is an isomorphism of* $S^{-1}A$ *modules.*

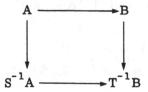

Proof. Let us define a map

$$g: S^{-1}M \to S^{-1}A \otimes_A M \text{ by } \frac{m}{s} \mapsto \frac{1}{s} \otimes m .$$

This is well-defined, for if $\frac{m}{s} = \frac{m'}{s'}$, then $t(s'm - sm') = 0$ for some $t \in S$, and

$$\frac{1}{s} \otimes m - \frac{1}{s'} \otimes m' = \frac{ts'}{tss'} \otimes m - \frac{ts}{tss'} \otimes m'$$

$$= \frac{1}{tss'} \otimes t(s'm - sm')$$

$$= 0 .$$

We have also $f(g(\frac{m}{s})) = f(\frac{1}{s} \otimes m) = \frac{m}{s}$; and

$$g(f(\sum_{i=1}^{r} \frac{a_i}{s_i} \otimes m_i)) = g(\sum_{i=1}^{r} \frac{a_i m_i}{s_i})$$

$$= g\left(\frac{\sum_{i=1}^{r} s_1 \ldots \hat{s}_i \ldots s_r a_i m_i}{s_1 \ldots s_r}\right)$$

$$= \frac{1}{s_1 \ldots s_r} \otimes \sum_{i=1}^{r} s_1 \ldots \hat{s}_i \ldots s_r a_i m_i$$

$$= \sum_{i=1}^{r} \frac{a_i}{s_i} \otimes m_i .$$

Thus $g = f^{-1}$. □

3.1.3 (Proposition). Let A be a ring and $S \subseteq A$ be multiplicative. Then $S^{-1}A$ is a flat A-module.

Proof. Let $F: M \to N$ be an injective morphism of A-modules. We must show that $S^{-1}A \otimes_A M \xrightarrow{1_{S^{-1}A} \otimes f} S^{-1}A \otimes_A N$ is injective; or by 3.1.2 that $S^{-1}M \xrightarrow{\frac{m}{s} \mapsto \frac{f(m)}{s}} S^{-1}N$ is injective. Let $\frac{f(m)}{s} = 0$; then $tf(m) = 0$ for some $t \in S$; so $f(tm) = 0$ and $tm = 0$ and $\frac{m}{s} = 0$. □

For example $\mathbf{Q} = \text{tot}(\mathbf{Z})$ is a flat $\mathbf{Z}$-module.

If M is an A-algebra and $S \subseteq A$ is multiplicative, we make $S^{-1}M$ into an $S^{-1}A$-algebra (and an A-algebra) by defining

$$\frac{m}{s}\frac{m'}{s'} = \frac{mm'}{ss'}.$$

This $S^{-1}A$-algebra structure is the same as that got from the natural one on $S^{-1}A \otimes_A M$ using 3.1.2.

3.2 LOCALISATION

Let A be a ring and $\mathfrak{p} \in \text{spec}(A)$. Then $A \setminus \mathfrak{p}$ is multiplicative and we define

$$A_{\mathfrak{p}} = (A \setminus \mathfrak{p})^{-1}A$$

called the localisation of A at $\mathfrak{p}$.

Note that $A_{\mathfrak{p}}$ is a non-zero ring if A is.

(If A is integral we usually regard $A_{\mathfrak{p}}$ as a subring of the field of fractions of A.)

From 3.1.1 we have an increasing map:

$$X = \{ \mathfrak{q} \in \operatorname{spec}(A) : \mathfrak{q} \subseteq \mathfrak{p} \} \xrightarrow{\text{bij}} \operatorname{spec}(A_{\mathfrak{p}}).$$

But X has a greatest member, $\mathfrak{p}$; so $A_{\mathfrak{p}}$ has a greatest proper ideal

$$\begin{aligned} \mathfrak{m}(A_{\mathfrak{p}}) &= \{ \tfrac{p}{s} : p \in \mathfrak{p} \text{ and } s \in A \setminus \mathfrak{p} \} \\ &= \mathfrak{p} A_{\mathfrak{p}} \\ &= (A \setminus \mathfrak{p})^{-1} \mathfrak{p} \end{aligned}$$

and is a local ring.

In a similar way we define the $A_{\mathfrak{p}}$-module $M_{\mathfrak{p}}$, isomorphic to $A_{\mathfrak{p}} \otimes_A M$, for any A-module M.

3.2.1 (Theorem). $\bigoplus_{\mathfrak{m}} A_{\mathfrak{m}}$ (taken over all maximal ideals $\mathfrak{m}$ of A) is a faithfully flat A-module.

Proof. By 3.1.3 each $A_{\mathfrak{m}}$ is flat, so $\bigoplus_{\mathfrak{m}} A_{\mathfrak{m}}$ is flat; and if $\mathfrak{n}$ is a maximal ideal of A, we have

$$\mathfrak{n}(\bigoplus_{\mathfrak{m}} A_{\mathfrak{m}}) \subseteq \bigoplus_{\mathfrak{m}} \mathfrak{n} A_{\mathfrak{m}} \subset \bigoplus_{\mathfrak{m}} A_{\mathfrak{m}} \quad \text{since}$$

$$\mathfrak{n} A_{\mathfrak{n}} = \mathfrak{m}(A_{\mathfrak{n}}) \subset A_{\mathfrak{n}}.$$

Thus by 2.3.1 $\bigoplus_{\mathfrak{m}} A_{\mathfrak{m}}$ is a faithfully flat A-module. □

3.2.1.1 (Corollary). *The sequence* $L \to M \to N$ *of* A-*modules is exact iff the sequence* $L_{\mathfrak{m}} \to M_{\mathfrak{m}} \to N_{\mathfrak{m}}$ *of* $A_{\mathfrak{m}}$-*modules is exact for every maximal ideal* $\mathfrak{m}$ *of* A.

Proof. $L \to M \to N$ is exact iff $(\bigoplus_{\mathfrak{m}} A_{\mathfrak{m}}) \otimes_A L \to (\bigoplus_{\mathfrak{m}} A_{\mathfrak{m}}) \otimes_A M \to (\bigoplus_{\mathfrak{m}} A_{\mathfrak{m}}) \otimes_A N$ is exact iff $\bigoplus_{\mathfrak{m}} (A_{\mathfrak{m}} \otimes_A L) \to \bigoplus_{\mathfrak{m}} (A_{\mathfrak{m}} \otimes_A M) \to \bigoplus_{\mathfrak{m}} (A_{\mathfrak{m}} \otimes_A N)$ is exact iff $L_{\mathfrak{m}} \to M_{\mathfrak{m}} \to N_{\mathfrak{m}}$ is exact for every maximal ideal $\mathfrak{m}$ of A. □

3.2.1.2 (Corollary) *Let* E *be an* A-*module. Then* E *is flat iff* $E_{\mathfrak{m}}$ *is a flat* $A_{\mathfrak{m}}$-*module for every maximal ideal* $\mathfrak{m}$ *of* A.

Proof. Suppose E is flat and let $M \to N$ be an injective morphism of $A_{\mathfrak{m}}$-modules. We have the commutative diagram

$$\begin{array}{ccc} E \otimes_A M & \xrightarrow{\text{inj}} & E \otimes_A N \\ \text{bij} \downarrow & & \downarrow \text{bij} \\ E_{\mathfrak{m}} \otimes_{A_{\mathfrak{m}}} M & \to & E_{\mathfrak{m}} \otimes_{A_{\mathfrak{m}}} N \end{array}$$

Where the vertical map is $e \otimes m \mapsto \frac{e}{1} \otimes m$ (and its inverse is $\frac{e}{s} \otimes m \mapsto e \otimes \frac{1}{s} m$). Thus $E_{\mathfrak{m}} \otimes_{A_{\mathfrak{m}}} M \xrightarrow{\text{inj}} E_{\mathfrak{m}} \otimes_{A_{\mathfrak{m}}} N$ and $E_{\mathfrak{m}}$ is a flat $A_{\mathfrak{m}}$-module.

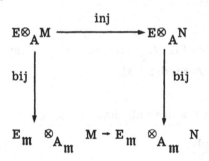

Conversely suppose that $E_{\mathfrak{m}}$ is flat for each $\mathfrak{m}$ and let $M \to N$ be an injective morphism of A-modules. Since $A_{\mathfrak{m}}$ is a flat A-module, we have $M_{\mathfrak{m}} \xrightarrow{\text{inj}} N_{\mathfrak{m}}$. Thus the diagram

$$\begin{array}{ccc} E_{\mathfrak{m}} \otimes_{A_{\mathfrak{m}}} M_{\mathfrak{m}} & \xrightarrow{\quad\text{inj}\quad} & E_{\mathfrak{m}} \otimes_{A_{\mathfrak{m}}} N_{\mathfrak{m}} \\ \Big\downarrow \text{bij} & & \Big\downarrow \text{bij} \\ A_{\mathfrak{m}} \otimes_A (E \otimes_A M) & \xrightarrow{\qquad\quad} & A_{\mathfrak{m}} \otimes_A (E \otimes_A N) \end{array}$$

commutes, where the vertical map is $\frac{e}{s} \otimes \frac{m}{t} \mapsto \frac{1}{st} \otimes (e \otimes m)$ (and its inverse is $\frac{a}{s} \otimes (e \otimes m) \mapsto \frac{ae}{s} \otimes m$). Therefore $(E \otimes_A M)_{\mathfrak{m}} \xrightarrow{\text{inj}} (E \otimes_A N)_{\mathfrak{m}}$ for every $\mathfrak{m}$, and by 3.2.1.1 $E \otimes_A M \xrightarrow{\text{inj}} E \otimes_A N$ and E is flat. □

3.2.1.3 (Corollary). Let A be an integral ring and k be its field of fractions, so that we can identify $A_{\mathfrak{m}}$ with a subring of k for any maximal ideal $\mathfrak{m}$ of A. Then $A = \bigcap_{\mathfrak{m}} A_{\mathfrak{m}}$ taken over all maximal ideals $\mathfrak{m}$ of A.

Proof. We can identify $k_{\mathfrak{m}}$ with k by the isomorphism: $A_{\mathfrak{m}} \otimes_A k \to k$ given by $\frac{a}{s} \otimes x \mapsto s^{-1}ax$ (its inverse is $y \mapsto \frac{1}{1} \otimes y$).

Let $B = \bigcap_{\mathfrak{m}} A_{\mathfrak{m}}$. Thus $A \subseteq B \subseteq A_{\mathfrak{m}} \subseteq k$ for all $\mathfrak{m}$; so $A_{\mathfrak{m}} \subseteq B_{\mathfrak{m}} \subseteq (A_{\mathfrak{m}})_{\mathfrak{m}} \subseteq k_{\mathfrak{m}}$ and $A_{\mathfrak{m}} \subseteq B_{\mathfrak{m}} \subseteq A_{\mathfrak{m}} \subseteq k$ by the above identification. Thus $A_{\mathfrak{m}} = B_{\mathfrak{m}}$ for all $\mathfrak{m}$ and by 3.2.1.1 the inclusion: $A \to B$ is onto. Thus $A = B$. □

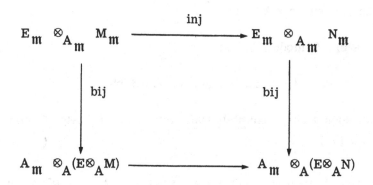

Exercise (hard). Let A be a ring. Show that the following statements are equivalent:

(i) every principal ideal of A is generated by an idempotent (i. e. an element $e \in A$ with $e^2 = e$)

(ii) every A-module is flat

(iii) $\mathfrak{n}(A) = \{0\}$ and spec(A) is Hausdorff.

(We call such a ring <u>absolutely flat.</u> See Bourbaki, Alg. Comm. I, §2, Ex 17.)

Let A be a ring and $f \in A$. We write A_f for $S^{-1}A$ where $S = \{f^n : n \in \omega\}$. The image of $\mathrm{spec}(A_f)$ in spec(A) is the open set $D(f) = \{\mathfrak{p} \in \mathrm{spec}(A) : f \notin \mathfrak{p}\}$; and the sets D(f) form a base for the Zariski topology, for

$$\mathrm{spec}(A) \setminus V(\mathfrak{a}) = \bigcup_{f \in \mathfrak{a}} D(f)$$

for any ideal $\mathfrak{a}$ of A.

Similarly we define M_f, etc.

If M is an A-module of finite type and $M_{\mathfrak{p}} = 0$ for some $\mathfrak{p} \in \mathrm{spec}(A)$, then $M_f = 0$ for some $f \in A \setminus \mathfrak{p}$: for if $(m_i)_{1 \le i \le n}$ generates M, we have $s_i m_i = 0$ for some $s_i \in A \setminus \mathfrak{p}$, and we let $f = s_1 \ldots s_n$. Conversely if $M_f = 0$ and $\mathfrak{p} \in D(f)$, $M_{\mathfrak{p}} = 0$.

<u>3.2.2 (Lemma).</u> <u>Let</u> A <u>be a ring and</u> $M \to N$ <u>be a morphism of</u> A-<u>modules.</u> <u>Let</u> $\mathfrak{p} \in \mathrm{spec}(A)$.

(i) <u>If</u> $M_{\mathfrak{p}} \xrightarrow{\text{onto}} N_{\mathfrak{p}}$ <u>and</u> N <u>is of finite type, then</u> $M_f \xrightarrow{\text{onto}} N_f$ <u>for some</u> $f \in A \setminus \mathfrak{p}$.

(ii) <u>If</u> $M_{\mathfrak{p}} \xrightarrow{\text{bij}} N_{\mathfrak{p}}$ <u>and</u> M <u>is of finite type and</u> N <u>is of finite presentation, then</u> $M_g \xrightarrow{\text{bij}} N_g$ <u>for some</u> $g \in A \setminus \mathfrak{p}$.

Proof. Let $0 \to R \to M \to N$ and $M \to N \to Q \to 0$ be exact.

(i) Then $M_{\mathfrak{p}} \xrightarrow{\text{onto}} N_{\mathfrak{p}} \to Q_{\mathfrak{p}} \to 0$ is exact and $Q_{\mathfrak{p}} = 0$; so $Q_f = 0$ for some $f \in A \setminus \mathfrak{p}$ since Q is of finite type; and $M_f \to N_f \to Q_f \to 0$ is exact: so $M_f \xrightarrow{\text{onto}} N_f$.

(ii) We have $0 \to R_f \to M_f \to N_f \to 0$ exact; so (since A_f is a flat A-algebra, and by 2.3.5) N_f is of finite presentation and M_f is of finite type and by 2.3.2 R_f is of finite type. However $0 \to R_{\mathfrak{p}} \to M_{\mathfrak{p}} \xrightarrow{\text{bij}} N_{\mathfrak{p}}$ is exact; so $R_{\mathfrak{p}} = 0$ and $(R_f)_{\mathfrak{p} A_f} = 0$. Thus $\frac{h}{1} R_f = 0$ for some $h \in A \setminus \mathfrak{p}$; and we take $g = fh$ to find that $R_g = 0$ and $Q_g = 0$: so that $M_g \xrightarrow{\text{bij}} N_g$. □

3.2.3 (Lemma). <u>Let</u> $(f_i)_{i \in I}$ <u>be a finite family in a ring</u> A <u>which generates the ideal</u> A. <u>Then</u> $\prod_{i \in I} A_{f_i}$ <u>is a faithfully flat</u> A-<u>algebra.</u>

Proof. Clearly $\prod_{i \in I} A_{f_i}$ is flat. Let $\mathfrak{m}$ be a maximal ideal of A. Then $f_i \notin \mathfrak{m}$ for some $i \in I$, so that $\mathfrak{m} A_{f_i} \subset A_{f_i}$ and $\prod_{i \in I} A_{f_i} \subset \prod_{i \in I} A_{f_i}$. Thus $\prod_{i \in I} A_{f_i}$ is faithfully flat by 2.3.1 (iii). □

3.3 PROJECTIVE MODULES AND LOCALISATION

We say that an A-module M is of countable type iff there is a family $(m_n)_{n \in \omega}$ which generates M. For example, modules of finite type are of countable type.

If M is of finite (resp. countable) type and N is a submodule of M, then M/N is of finite (resp. countable) type. Conversely if N and M/N are of finite (resp. countable) type, so is M. Thus if $M = P \oplus Q$, then M is of finite (resp. countable) type iff both P and Q are so too. If $(M_n)_{n \in \omega}$ is a family of modules of countable type, $\bigoplus_{n \in \omega} M_n$ is also of countable type.

3.3.1 (Lemma). (i) Let M be an A-module and P, Q, Q' be submodules such that $M = P \oplus Q = P \oplus Q'$. Then there is an isomorphism: $Q \to Q'$.

(ii) Let N be an A-module and let Q, L and M be submodules with $L \subseteq M$ and $N = L \oplus Q$. Then $M = L \oplus (Q \cap M)$.

(iii) Let $f: M \to M$ be an A-module morphism with $f = f^2$. Then $M = f[M] \oplus (1_M - f)[M]$.

Proof. (i) It is $\{(x, x'): x \in Q;\ x' \in Q';\ x - x' \in P\}$.

(ii) and (iii) are immediate. □

Let us recall some facts about ordinals.

A set α is called ordinal iff $\alpha \subseteq \mathscr{P}(\alpha)$ and $\in|\alpha$ is a strict well order on α. For example $\emptyset$ is an ordinal; and if α is an ordinal and $\beta \in \alpha$, then β is an ordinal. If α is an ordinal, $\alpha \notin \alpha$ and the successor $\mathrm{suc}(\alpha) = \alpha \cup \{\alpha\}$ of α is an ordinal distinct from α. Not every ordinal is a successor: in fact an ordinal α is not a successor iff $\alpha = \bigcup(\alpha)$ iff $\mathrm{suc}(\beta) \in \alpha$

for all $\beta \in \alpha$. Such an α we call a limit ordinal: for example $\emptyset$ and ω.

The class On of all ordinals is well ordered by inclusion. If $X \subseteq$ On, and $\alpha \subseteq X$ implies $\alpha \in X$ for all $\alpha \in$ On, then $X =$ On. (Transfinite induction.)

If f is a function whose domain is the universe, there is one and only one function g whose domain is On and for which $g(\alpha) = f(g|\alpha)$ for all $\alpha \in$ On. (Transfinite recursion.)

The axiom of substitution guarantees that if A is a set, then no map: On $\rightarrow$ A can be injective (for otherwise, On would be a set and On $\in$ On, a contradiction).

Exercise (for those unfamiliar with ordinals). Show that if α and β are ordinals, then $\alpha \subseteq \beta$ or $\beta \subseteq \alpha$. Use this to prove all the above statements.

3.3.2 **(Lemma)** (Kaplansky). Let P be a projective A-module. Then $P = \bigoplus_{\gamma \in \alpha} Q_\gamma$ for a family $(Q_\gamma)_{\gamma \in \alpha}$ of submodules Q_γ of countable type.

Proof. Let $P \oplus Q = M$ a free module and $p: M \rightarrow M$ be the projection onto P, so that $p = p^2$. Plainly $M = \bigoplus_{i \in I} M_i$ for some family $(M_i)_{i \in I}$ of submodules of countable type. Let $(\chi_i: M \rightarrow M_i)_{i \in I}$ be the associated family of projections.

Define a relation $R = \{(i, j) \in I \times I: \chi_j(p(x)) \neq 0$ for some $x \in M_i\}$ on I, so that for all $i \in I$ the set $R[\{i\}] = \{j \in I: (i, j) \in R\}$ is countable. For $n \in \omega$ write R^n for $R \circ R \circ \ldots \circ R$ (n times).

Let us define by transfinite recursion the function $\beta \mapsto I_\beta$ taking On to $\mathscr{P}(I)$ as follows:

(i) if β is a limit ordinal let $I_\beta = \bigcup_{\gamma \in \beta} I_\gamma$

(ii) if $\beta = \mathrm{suc}(\gamma)$ for some γ and $I_\gamma = I$, let $I_\beta = I$

(iii) if $\beta = \mathrm{suc}(\gamma)$ for some γ and $I_\gamma \subset I$, choose $i \in I \backslash I_\gamma$ and let $I_\beta = I_\gamma \cup (\bigcup_{n \in \omega} R^n[\{i\}])$.

Then (by transfinite induction where necessary) we have

(i) I_β increases with β

(ii) $I_\emptyset = \emptyset$

(iii) $I_{\mathrm{suc}(\beta)} \backslash I_\beta$ is countable for all β

(iv) if $i \in I_\beta$ and $(i, j) \in R$, then $j \in I_\beta$.

Suppose $I_\alpha \subset I$ for all α; then $\mathrm{On} \to \mathscr{P}(I)$ is strictly increasing, contradicting the axiom of substitution. Thus $I_\alpha = I$ for some least limit ordinal α, and $I = \bigcup_{\beta \in \alpha} I_\beta$.

Define the family $(E_\beta)_{\beta \in \alpha}$ of submodules of M as follows:

$$E_\beta = \bigoplus_{i \in I_\beta} M_i$$

so that $M = \bigcup_{\beta \in \alpha} E_\beta$. Also $p[E_\beta] \subseteq E_\beta$: for if $x \in M_i$ and $i \in I_\beta$, then $(i, j) \notin R$ implies $\chi_j(p(x)) = 0$, so that

$$p(x) \in \bigoplus_{(i,j) \in R} M_j \subseteq \bigoplus_{j \in I_\beta} M_j \quad \text{by (iv)}$$

$$= E_\beta .$$

We define too $(F_\beta)_{\beta \in \alpha}$ by:

$$F_\beta = \bigoplus_{i \in I_{suc(\beta)} \setminus I_\beta} M_i \ ;$$

so that $E_{suc(\beta)} = E_\beta \oplus F_\beta$ for all $\beta \in \alpha$; and by (iii) F_β is of countable type for all $\beta \in \alpha$.

For $\beta \in \alpha$ we have

$$\begin{aligned} E_{suc(\beta)} &= E_\beta \oplus F_\beta \\ &= p[E_\beta] \oplus (1 - p)[E_\beta] \oplus F_\beta \quad \text{by } 3.3.1 \text{ (iii)} \end{aligned}$$

so that $p[E_{suc(\beta)}] = p[E_\beta] \oplus Q_\beta$ by 3.3.1 (ii) where $Q_\beta = p[E_{suc(\beta)}] \cap ((1 - p)[E_\beta] \oplus F_\beta)$. Also $E_{suc(\beta)} = p[E_{suc(\beta)}] \oplus (1 - p)[E_{suc(\beta)}]$; and (putting $1 - p$ in place of p)

$$(1 - p)[E_{suc(\beta)}] = (1 - p)[E_\beta] \oplus L_\beta \quad \text{for some } L_\beta .$$

Therefore

$$E_{suc(\beta)} = p[E_\beta] \oplus (1 - p)[E_\beta] \oplus Q_\beta \oplus L_\beta$$

and by 3.3.1 (i) $Q_\beta \oplus L_\beta$ and F_β are isomorphic. Thus Q_β is of countable type for all $\beta \in \alpha$.

Let us show by transfinite induction that $p[E_\beta] = \bigoplus_{\gamma \in \beta} Q_\gamma$ for all $\beta \in \alpha$. If β is a limit ordinal, then

$$p[E_\beta] = \bigcup_{\gamma \in \beta} p[E_\gamma] = \bigcup_{\gamma \in \beta} \bigoplus_{\delta \in \gamma} Q_\delta = \bigoplus_{\delta \in \beta} Q_\delta \ ;$$

and if $\beta = suc(\gamma)$ then

$$p[E_\beta] = p[E_\gamma] \oplus Q_\gamma = \bigoplus_{\delta \in \beta} Q_\delta \ .$$

Thus $p[E_\beta] = \bigoplus_{\gamma \in \beta} Q_\gamma$ for all $\beta \in \alpha$; and

$$P = p[M] = \bigcup_{\beta \in \alpha} p[E_\beta] = \bigcup_{\beta \in \alpha} \bigoplus_{\gamma \in \beta} Q_\gamma = \bigoplus_{\gamma \in \alpha} Q_\gamma \ . \square$$

3.3.3 **(Lemma).** Let A be a local ring; P be a projective A-module; and $x \in P$ with $x \neq 0$. Then there exist submodules Q, R of P such that

(i) $x \in Q$

(ii) Q is free and of finite type

(iii) $P = Q \oplus R$.

Proof. Let $P \oplus L = M$, a free module. If $(m_i)_{i \in I}$ is a base of M and $x = \sum_{i \in I} \xi_i m_i$ we define $J = \{i \in I: \xi_i \neq 0\}$. Among all bases choose one for which the finite set J has the least possible number of elements. Then if $j \in J$ we have $\xi_j \notin \sum_{i \neq j} A\xi_i$: for if $\xi_j = \sum_{i \neq j} a_i \xi_i$, define

$$m_i' = m_i + a_i m_j \quad \text{if } i \in J \text{ and } i \neq j$$

$$= m_i \quad \text{if } i = j \text{ or } i \in I \backslash J$$

so that $(m_i')_{i \in I}$ bases M and $x = \sum_{i \neq j} \xi_i m_i'$, a contradiction.

Now for each $i \in I$ let $m_i = y_i + z_i$ for $y_i \in P$ and $z_i \in L$. Thus $x = \sum_{i \in J} \xi_i y_i$ and $\sum_{i \in J} \xi_i z_i = 0$. Let $z_i = \sum_{j \in I} \eta_{ij} m_j$ for each $i \in J$; then $\sum_{j \in I} (\sum_{i \in J} \eta_{ij} \xi_i) m_j = 0$ and $\sum_{i \in J} \eta_{ij} \xi_i = 0$ for all $j \in I$. Thus $\eta_{ij} \in \mathfrak{m}(A)$ for all $i \in J$ and $j \in I$: for if $\eta_{ij} \in \mathfrak{u}(A)$, then $\xi_i \in \sum_{j \neq i} A\xi_j$. But $y_i = m_i - \sum_{j \in I} \eta_{ij} m_j$ for $i \in J$: so if we define

$$m_i'' = y_i \quad \text{if } i \in J$$

$$\quad = m_i \quad \text{if } i \in I \backslash J$$

we see that $(m_i'')_{i \in I}$ bases M; and we write

$$Q = \sum_{i \in J} Ay_i$$

$$R = P \cap (\sum_{i \in I \backslash J} Am_i) .$$

Thus

(i) $x \in Q$

(ii) Q is free and of finite type, and

(iii) $P = Q \oplus R$ by 3.3.1 (ii) □

<u>3.3.4 (Theorem).</u> <u>Let</u> A <u>be a local ring and</u> P <u>be a projective</u> A-<u>module.</u> <u>Then</u> P <u>is free.</u>

Proof. By 3.3.2 we may suppose that a family $(w_\gamma)_{\gamma \in \omega}$ generates P.

Let us define recursively sequences $(Q_n)_{n \in \omega}$, $(P_n)_{n \in \omega}$, $(p_n)_{n \in \omega}$ as follows:

(i) $Q_0 = 0$; $P_0 = P$; $p_0 = 1_P$.

(ii) If $p_n(w_\gamma) = 0$ for all $\gamma \in \omega$ we let $P_{n+1} = Q_{n+1} = 0$ and $p_{n+1} = 0_P$.

(iii) If $p_n(w_\gamma) \neq 0$ for some least $\gamma \in \omega$ we use 3.3.3 to choose submodules P_{n+1} and Q_{n+1} of P_n such that

(a) $p_n(w_\gamma) \in Q_{n+1}$

(b) Q_{n+1} is free of finite type

(c) $P_n = Q_{n+1} \oplus P_{n+1}$

and define $p_{n+1}: P \to P_{n+1}$ to be the projection.

Thus the sum $\sum_{n \in \omega} Q_n$ is direct; and if $w_\gamma \notin \bigoplus_{n \in \omega} Q_n$ for some least $\gamma \in \omega$, we have $w_1, \ldots, w_{\gamma-1} \in \bigoplus_{\substack{n \in \omega \\ n \leq N}} Q_n$ for some least $N \in \omega$; thus $p_N(w_\gamma) \neq 0$ so that $w_\gamma \in \bigoplus_{\substack{n \in \omega \\ n \leq N+1}} Q_n$.

Thus $P = \bigoplus_{n \in \omega} Q_n$ and is free. □

(Without the rather difficult 3.3.2 we see that projective modules of countable type over local rings are free.)

3.3.5 (Lemma). Let A be a local ring and M be an A-module of finite type. Let $\mathfrak{m} = \mathfrak{m}(A)$. Suppose $A/\mathfrak{m} \otimes_A M = 0$. Then $M = 0$.

Proof. We have $A/\mathfrak{m} \otimes_A M \xrightarrow{\text{iso}} M/\mathfrak{m} M$ by $(a + \mathfrak{m}) \otimes m \mapsto am + \mathfrak{m} M$. Thus $\mathfrak{m} M = M$. Let $(m_i)_{1 \leq i \leq n}$ generate M for some least $n \in \omega$. If $n > 0$ we have $m_n = \sum_{i=1}^{n} \lambda_i m_i$ for some $\lambda_i \in \mathfrak{m}$, so that $m_n = (1 - \lambda_n)^{-1} \sum_{i=1}^{n-1} \lambda_i m_i$ and $(m_i)_{1 \leq i \leq n-1}$ generates M, a contradiction. Thus $n = 0$ and $M = 0$. □

(This is a special case of a famous lemma of Nakayama.)

3.3.5a (Lemma). Let $f: M \xrightarrow{\text{onto}} N$ be an A-module morphism; and let $(x_i)_{1 \le i \le m}$ generate M and $(y_j)_{1 \le j \le n}$ base N, where $m \le n$. Then f is an isomorphism.

Proof. We may suppose that $m = n$.

There are matrices $(a_{ij})_{1 \le i, j \le n}$ and $(b_{ki})_{1 \le k, i \le n}$ over A such that $f(x_i) = \sum_{j=1}^{n} a_{ij} y_j$, $y_k = \sum_{i=1}^{n} b_{ki} f(x_i)$ for $i, k = 1, \ldots, n$. Thus

$$y_k = \sum_{j=1}^{n} \left(\sum_{i=1}^{n} b_{ki} a_{ij} \right) y_j$$

so that
$$\sum_{i=1}^{n} b_{ki} a_{ij} = 1 \text{ if } k = j$$
$$= 0 \text{ if } k \ne j .$$

That is, $(b_{ki})_{k,i}$ is the inverse of $(a_{ij})_{i,j}$. Suppose $f(\sum_{i=1}^{n} \xi_i x_i) = 0$ for $\xi_i \in A$. Then $\sum_{j=1}^{n} (\sum_{i=1}^{n} \xi_i a_{ij}) y_j = 0$ so that $(\xi_i)_i (a_{ij})_{i,j} = 0$. Postmultiplying by $(b_{ki})_{k,i}$ we obtain $(\xi_i)_i = 0$, so that $\sum_{i=1}^{n} \xi_i x_i = 0$ and f is injective. □

(This proof was contributed by Barnard.)

3.3.6 (Theorem). Let A be a local ring and M be an A-module of finite presentation. The following conditions are equivalent:

(i) M <u>is free</u>

(ii) M <u>is projective</u>

(iii) M <u>is flat</u>

(iv) $\mathfrak{m}(A)\otimes_A M \xrightarrow{\text{inj}} M$.

Proof. Suppose (iv) holds and let $\mathfrak{m} = \mathfrak{m}(A)$ and $k = A/\mathfrak{m}$. Then $k\otimes_A M$ is a k-module of finite type: say $(1\otimes m_i)_{i\in I}$ bases $k\otimes_A M$. Let $A^I \to M$ by $(\xi_i)_{i\in I} \mapsto \sum_{i\in I} \xi_i m_i$ and $k^I \xrightarrow{\text{iso}} k\otimes_A M$ by $(\eta_i)_{i\in I} \mapsto \sum_{i\in I} \eta_i \otimes m_i$. Then

$$\begin{array}{ccc} k\otimes_A A^I & \xrightarrow[\text{iso}]{\text{nat}} & k^I \\ & \searrow & \downarrow{\scriptstyle\text{iso}} \\ & & k\otimes_A M \end{array}$$

commutes: so $k\otimes_A A^I \xrightarrow{\text{iso}} k\otimes_A M$.

Suppose $A^I \to M \to Q \to 0$ is exact: then $k\otimes_A A^I \xrightarrow{\text{iso}} k\otimes_A M \to k\otimes_A Q \to 0$ is exact and $k\otimes_A Q = 0$. But Q is of finite type: so $Q = 0$ by 3.3.5, and $A^I \xrightarrow{\text{onto}} M$.

Suppose $0 \to R \to A^I \to M \to 0$ is exact: we have a commutative diagram

$$\begin{array}{ccccccccc} & & & & & & 0 & & \\ & & & & & & \downarrow & & \\ & & \mathfrak{m}\otimes_A R & \longrightarrow & \mathfrak{m}\otimes_A A^I & \longrightarrow & \mathfrak{m}\otimes_A M & & \\ & & \downarrow & & \downarrow & & \downarrow & & \\ 0 & \longrightarrow & R & \longrightarrow & A^I & \longrightarrow & M & \longrightarrow & 0 \\ & & \downarrow & & \downarrow & & \downarrow & & \\ & & k\otimes_A R & \longrightarrow & k\otimes_A A^I & \xrightarrow{\text{iso}} & k\otimes_A M & & \\ & & \downarrow & & & & & & \\ & & 0 & & & & & & \end{array}$$

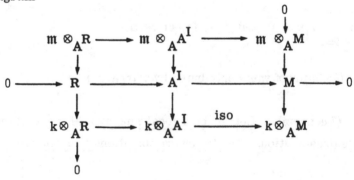

with exact rows and columns. Let $x \in k \otimes_A R$. Then $x \mapsto 0 \in k \otimes_A A^I$. Let $y \mapsto x$ for $y \in R$, and $y \mapsto z \in A^I$. Then $z \mapsto 0 \in k \otimes_A A^I$ so $w \mapsto z$ for some $w \in \mathfrak{m} \otimes_A A^I$. But $z \mapsto 0 \in M$ so $w \mapsto 0 \in M$ and $w \mapsto 0 \in \mathfrak{m} \otimes_A M$. Thus $u \mapsto w$ for some $u \in \mathfrak{m} \otimes_A R$; and $u \mapsto z \in A^I$; so $u \mapsto y \in R$ and $y \mapsto 0 = x \in k \otimes_A R$. That is, $k \otimes_A R = 0$. But by 2.3.2, R is of finite type. Thus by 3.3.5, $R = 0$: that is, $A^I \xrightarrow{\text{iso}} M$ and (i) holds.

The other implications are immediate. □

<u>3.3.7 (Theorem). Let A be a ring and P be an A-module. The following conditions are equivalent:</u>

(i) <u>P is projective and of finite type</u>

(ii) <u>P is of finite presentation; and $P_{\mathfrak{m}}$ is a free $A_{\mathfrak{m}}$-module for every maximal ideal $\mathfrak{m}$ of A</u>

(iii) <u>P is of finite type; $P_{\mathfrak{p}}$ is a free $A_{\mathfrak{p}}$-module for every $\mathfrak{p} \in \text{spec}(A)$, of rank $r_{\mathfrak{p}}$ say; and $\mathfrak{p} \mapsto r_{\mathfrak{p}}$ is continuous</u>

(iv) <u>there is a finite family $(f_i)_{i \in I}$ in A, generating the ideal A, and such that for all $i \in I$ the A_{f_i}-module P_{f_i} is free and of finite type.</u>

Proof. It is convenient to prove at the same time a fifth condition:

(v) for every maximal ideal $\mathfrak{m}$ of A there exists $f \in A \setminus \mathfrak{m}$ such that P_f is a free A_f-module of finite type.

(i) $\Longrightarrow$ (ii) Immediate from 3. 3. 4 and 2. 3. 5.

(ii) $\Longrightarrow$ (v) Let $\mathfrak{m}$ be maximal; then $(1 \otimes p_i)_{i \in I}$ bases $A_{\mathfrak{m}} \otimes_A P$, say. The map: $A^I \to P$ given by $(\xi_i)_{i \in I} \mapsto \sum_{i \in I} \xi_i p_i$ is such that $(A^I)_{\mathfrak{m}} \xrightarrow{\text{iso}} P_{\mathfrak{m}}$. Thus by 3. 2. 2 (ii), $(A^I)_f \xrightarrow{\text{iso}} P_f$ for some $f \in A \setminus \mathfrak{m}$; and $(A^I)_f \xrightarrow{\text{iso}} (A_f)^I$: that is, P_f is free and of finite type.

(v) $\Longrightarrow$ (iv) Let $X = \{f \in A: P_f$ is a free A_f-module of finite type$\}$. Then $X \not\subseteq \mathfrak{m}$ for every maximal $\mathfrak{m}$; so X generates the ideal A and $1 = \sum_{i \in I} a_i f_i$ for some $a_i \in A$; $f_i \in X$; and I finite.

(iv) $\Longrightarrow$ (iii) The $\prod_{i \in I} A_{f_i}$-module $\prod_{i \in I} P_{f_i}$ is of finite type; and we have $\prod_{i \in I} P_{f_i} \xrightarrow{\text{iso}} (\prod_{i \in I} A_{f_i}) \otimes_A P$ by:

$$\left(\frac{p_i}{f_i^{n_i}}\right)_{i \in I} \longmapsto \sum_{i \in I} g_i \otimes p_i \quad \text{where:}$$

$$g_i(j) = \frac{1}{f_i^{n_i}} \quad \text{if } i = j$$

$$= \frac{0}{1} \quad \text{if } i \neq j .$$

Thus P is of finite type by 3. 2. 3 and 2. 3. 5.

For each $\mathfrak{p} \in \operatorname{spec}(A)$ there exists $i \in I$ such that $\mathfrak{p} \in D(f_i)$; so that $(A_{f_i})_{\mathfrak{p} A_{f_i}} \xrightarrow{\text{iso}} A_{\mathfrak{p}}$ by:

$$\frac{a}{f_i^n} \bigg/ \frac{s}{f_i^m} \longmapsto \frac{f_i^m a}{s f_i^n}$$

for $s \in A \setminus \mathfrak{p}$ and $a \in A$ (the inverse is $\frac{a}{s} \mapsto \frac{a}{1} / \frac{s}{1}$). Similarly $(P_{f_i})_{\mathfrak{p} A_{f_i}} \xrightarrow{\text{iso}} P_{\mathfrak{p}}$ so that $P_{\mathfrak{p}}$ and P_{f_i} are free $A_{\mathfrak{p}}$ - and A_{f_i} -modules of finite type and of the same rank by 1. 3. 1. Further $r_{\mathfrak{p}} = n$ iff $\mathfrak{p} \in \cup D(f_i)$ taken over $i \in I$ such that P_{f_i} is of rank n; so $\mathfrak{p} \mapsto r_{\mathfrak{p}}$ is continuous.

(iii) $\Longrightarrow$(v) Let $\mathfrak{m}$ be maximal and $(\frac{p_i}{1})_{i \in I}$ base $P_{\mathfrak{m}}$, where I has n elements. Map $A^I \to P$ by $(\xi_i)_{i \in I} \mapsto \sum_{i \in I} \xi_i p_i$. Thus $(A^I)_{\mathfrak{m}} \xrightarrow{\text{onto}} P_{\mathfrak{m}}$ and $(A^I)_g \xrightarrow{\text{onto}} P_g$ for some $g \in A \setminus \mathfrak{m}$ by 3. 2. 2 (i). By the continuity there is an $h \in A \setminus \mathfrak{m}$ such that if $\mathfrak{p} \in D(h)$, then $r_{\mathfrak{p}} = n$. Let $f = gh$. Then $f \in A \setminus \mathfrak{m}$ and $(A^I)_f \xrightarrow{\text{onto}} P_f$ still, and $r_{\mathfrak{p}} = n$ for all $\mathfrak{p} \in D(f)$. Thus $(A^I)_{\mathfrak{p}} \xrightarrow{\text{onto}} P_{\mathfrak{p}}$ for all $\mathfrak{p} \in D(f)$; and $(A^I)_{\mathfrak{p}}$ (isomorphic to $(A_{\mathfrak{p}})^I$) and $P_{\mathfrak{p}}$ are free of rank n; so $(A^I)_{\mathfrak{p}} \xrightarrow{\text{iso}} P_{\mathfrak{p}}$ for all $\mathfrak{p} \in D(f)$ by 3. 3. 5a.

Let $\mathfrak{n}$ be a maximal ideal of A_f and $\mathfrak{q}$ be its image in spec(A), so that $\mathfrak{q} \in D(f)$. We have the commutative diagram

$$\begin{array}{ccc} ((A_f)^I)_{\mathfrak{n}} & \longrightarrow & (P_f)_{\mathfrak{n}} \\ \big\downarrow \text{iso} & & \big\downarrow \\ (A_{\mathfrak{q}})^I & \xrightarrow{\text{iso}} & P_{\mathfrak{q}} \end{array}$$

where the vertical map is given by

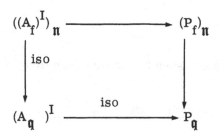

$$\frac{x}{f^n} / \frac{s}{f^m} \longmapsto \frac{f^m x}{sf^n}$$

for $s \in A \setminus \mathfrak{q}$; thus $((A_f)^I)_{\mathfrak{n}} \xrightarrow{\text{iso}} (P_f)_{\mathfrak{n}}$ for all maximal ideals $\mathfrak{n}$ of A_f, so that $(A_f)^I \xrightarrow{\text{iso}} P_f$ by 3.2.1.1.

(iv) $\Longrightarrow$ (i) By 3.2.3 and 2.3.5 it suffices to show that the $\prod_{i \in I} A_{f_i}$-module of finite type $\prod_{i \in I} P_{f_i}$ (isomorphic to $(\prod_{i \in I} A_{f_i}) \otimes_A P$) is projective. Let $(L_i)_{i \in I}$ be a family of A_{f_i}-modules of finite type such that each $P_{f_i} \oplus L_i$ is free and has the same rank. Let $\phi_i : P_{f_i} \to P_{f_i} \oplus L_i$ and $\psi_i : P_{f_i} \oplus L_i \to P_{f_i}$ be the natural A_{f_i}-module morphisms. Then $\prod_{i \in I} (P_{f_i} \oplus L_i)$ is clearly a free $\prod_{i \in I} A_{f_i}$-module and $(\prod_{i \in I} \psi_i) \circ (\prod_{i \in I} \phi_i) = 1_{(\prod_{i \in I} P_{f_i})}$ so that $\prod_{i \in I} P_{f_i}$ is projective (compare the note at the end of 2.1). □

3.3.7.1 **(Corollary).** A flat module of finite presentation is projective.

Proof. 3.3.6 and 3.3.7 (ii). □

(This can be proved ad hoc even for non-commutative rings.)

3.3.7.2 **(Corollary).** A flat module of finite type over a Noetherian ring is projective. □

If spec(A) is connected (for example if A is integral or local) and P is a projective A-module of finite type, the function $\mathfrak{p} \longmapsto r_{\mathfrak{p}}$ is constant.

If P is a projective A-module of finite type and $\mathfrak{p} \mapsto r_{\mathfrak{p}}$ is constant, we call its constant value the rank of P.

3.4 SUBMODULES OF FRACTION ALGEBRAS

3.4.1 (Theorem). *Let M be an A-module of finite type. The following two conditions are equivalent:*

(i) M *is projective of rank* 1

(ii) $M \otimes_A N$ *is isomorphic to* A *for some* A-*module* N.

Moreover if (ii) holds then N *is isomorphic to the dual* M^* *of* M.

Proof. Suppose (i) holds and let $M \otimes_A M^* \to A$ be the natural map. Let $\mathfrak{m}$ be a maximal ideal of A. Then $(M \otimes_A M^*)_{\mathfrak{m}} \xrightarrow{\text{iso}} M_{\mathfrak{m}} \otimes_{A_{\mathfrak{m}}} (M^*)_{\mathfrak{m}}$ by $\frac{m \otimes \phi}{s} \mapsto \frac{m}{s} \otimes \frac{\phi}{1}$ (whose inverse is $\frac{m}{s} \otimes \frac{\phi}{t} \mapsto \frac{m \otimes \phi}{st}$). However M is of finite presentation and $A_{\mathfrak{m}}$ is a flat A-algebra; thus by 2.3.4 $(M^*)_{\mathfrak{m}}$ is isomorphic to the dual $(M_{\mathfrak{m}})^*$ of the $A_{\mathfrak{m}}$-module $M_{\mathfrak{m}}$. Thus $M \otimes_A M^* \to A$ localises to the natural map: $M_{\mathfrak{m}} \otimes_{A_{\mathfrak{m}}} (M_{\mathfrak{m}})^* \to A_{\mathfrak{m}}$ but by 3.3.7 (ii) $M_{\mathfrak{m}}$ is free of rank 1; so $M_{\mathfrak{m}} \otimes_{A_{\mathfrak{m}}} (M_{\mathfrak{m}})^* \xrightarrow{\text{iso}} A_{\mathfrak{m}}$ and by 3.2.1.1 $M \otimes_A M^* \xrightarrow{\text{iso}} A$ and (ii) holds.

Suppose (ii) holds. By 3.3.7 (iii) it is enough to prove that M is free of rank 1 under the assumption that A is local, with $\mathfrak{m} = \mathfrak{m}(A)$ and $k = A/\mathfrak{m}$, say. Tensoring $M \otimes_A N \xrightarrow{\text{iso}} A$

by k we obtain $M/\mathfrak{m}M \otimes_k N/\mathfrak{m}N \xrightarrow{\text{iso}} k$; thus $M/\mathfrak{m}M$ has rank 1 as a k-module: say $m + \mathfrak{m}M$ bases $M/\mathfrak{m}M$. Let $Am \to M \to Q \to 0$ be exact: then $k \otimes_A Am \to k \otimes_A M \to k \otimes_A Q \to 0$ is exact; but the diagram

$$\begin{array}{ccc} k \otimes_A Am & \longrightarrow & k \otimes_A M \\ \Big\downarrow \text{iso} & & \Big\downarrow \text{iso} \\ Am/\mathfrak{m}m & \xrightarrow{\text{onto}} & M/\mathfrak{m}M \end{array}$$

commutes; and so $k \otimes_A Q = 0$. But Q is of finite type, and thus $Q = 0$ by 3.3.5 and $M = Am$. Thus $A \xrightarrow{\text{onto}} M$ by $\lambda \mapsto \lambda m$. Suppose $\lambda m = 0$; and let $1 \mapsto m \otimes n$ under $A \to M \otimes_A N$; so $\lambda \mapsto \lambda m \otimes n = 0$ and $\lambda = 0$. Thus $A \xrightarrow{\text{iso}} M$ and M is free of rank 1. Therefore (by 3.3.7 (iii) as already explained) (i) holds.

Finally the A-modules N, $A \otimes_A N$, $M \otimes_A M^* \otimes_A N$, $M \otimes_A N \otimes_A M^*$, $A \otimes_A M^*$ and M^* are isomorphic. □

It is almost as if projective modules of rank 1 formed a group under $\otimes_A$ with identity A, and inverse M^* of M. However they form a proper class C and not a set. Thus define the Picard group of A to be an abelian group $\mathrm{Pic}(A)$ and a function $[\ ]: C \to \mathrm{Pic}(A)$ such that

(i) M is isomorphic to $N \Rightarrow [M] = [N]$

(ii) $[M \otimes_A N] = [M] + [N]$

(iii) $\mathrm{Pic}(A)$ is universal for (i) and (ii).

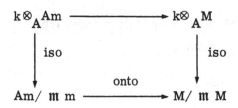

Exercise. Construct such a Pic(A).

For the rest of this section A is a ring; $S \subseteq A$ is multiplicative; $B = S^{-1}A$; and A' is the image of A in $S^{-1}A$.

If M and N are sub-A-modules of B we define the sub-A-modules

$$MN = \{ \sum_{i \in I} m_i n_i : I \text{ finite};\ m_i \in M;\ n_i \in N \}$$

$$(M:N) = \{ b \in B : bN \subseteq M \} \qquad \underline{\text{(ratio)}}$$

so that $A'M = M$ and $N(M:N) \subseteq M$.

3.4.2 (Lemma). Let M be a sub-A-module of B. Then the following conditions are equivalent:

(i) $\frac{s}{1} \in M$ for some $s \in S$

(ii) $BM = B$

(iii) $B \otimes_A M \xrightarrow{\text{bij}} B$ by $b \otimes m \mapsto bm$.

Proof. Suppose (i) holds and let $b \in B$. Then $b = (b\frac{1}{s})\frac{s}{1} \in BM$: that is, (ii) holds.

Suppose (ii) holds and let us prove (iii). Plainly $B \otimes_A M \xrightarrow{\text{onto}} B$. Since B is a flat A-module, $B \otimes_A M \xrightarrow{\text{inj}} B \otimes_A B$, and it remains to show that $B \otimes_A B \xrightarrow{\text{inj}} B$. Note first that

$$\frac{a}{s} \otimes \frac{1}{1} = \frac{1}{s} \otimes \frac{a}{1} = \frac{1}{s} \otimes s(\frac{a}{s}) = \frac{1}{1} \otimes \frac{a}{s};$$

so that

$$x \otimes y = (x \otimes 1)(1 \otimes y) = (1 \otimes x)(1 \otimes y) = 1 \otimes xy$$

in the A-algebra $B \otimes_A B$. Thus if $\sum_{i \in I} x_i y_i = 0$ for $x_i, y_i \in B$, we have

$$\sum_{i \in I} x_i \otimes y_i = \sum_{i \in I} 1 \otimes x_i y_i = 1 \otimes \sum_{i \in I} x_i y_i = 0$$

Finally suppose (iii) holds. Then $\frac{1}{1} = \sum_{i \in I} \frac{a_i}{s_i} m_i$ for some a_i, m_i, s_i; so $\frac{s_1 \dots s_n}{1} \in M$: that is, (i) holds.□

We call a submodule M of B which satisfies the conditions of 3.4.2 <u>non-degenerate.</u>

We have $M(A':M) \subseteq A'$ for any submodule M of B. The following conditions are equivalent:

(i) $M(A':M) = A'$

(ii) $\frac{1}{1} \in M(A':M)$

(iii) $MN = A'$ for some submodule N of B;

and if (iii) holds, $N = (A':M)$. We call such a submodule M <u>invertible</u>. If M is invertible, we have

$$BM = BBM \supseteq BM(A':M) = BA' = B$$

so that M is non-degenerate. Plainly the invertible submodules form an abelian group under multiplication, with identity A', and inverse (A':M) of M.

3.4.3 (Theorem). *Let* M *be a non-degenerate submodule of* B. *The following conditions are equivalent:*

(i) M *is invertible*

(ii) M *is a projective* A'-*module*

(iii) M *is a projective* A'-*module of rank* 1.

Proof. Suppose (i) holds. Then $\frac{1}{1} \in M(A':M)$ and $\frac{1}{1} = \sum_{i \in I} m_i n_i$ for some finite I and $m_i \in M$ and $n_i \in (A':M)$. Let $\phi : M \to A'^I$ by $m \longmapsto (n_i m)_{i \in I}$ and $\psi : A'^I \to M$ by $(\xi_i)_{i \in I} \longmapsto \sum_{i \in I} \xi_i m_i$. Then $\psi \circ \phi = 1_M$ and M is a projective A'-module of finite type. In particular M is a flat A'-module; and $M \otimes_A B \xrightarrow{\text{iso}} M \otimes_{A'} B$ by $m \otimes b \longmapsto m \otimes b$; so we have the commutative diagram

$$\begin{array}{ccc}
M \otimes_{A'} (A':M) & \xrightarrow{\text{onto}} & A' \\
\Big\downarrow \text{inj} & & \Big\downarrow \\
M \otimes_{A'} B & & \\
\Big\downarrow \text{iso} & & \\
M \otimes_A B & \xrightarrow{\text{iso}} & B
\end{array}$$

and $M \otimes_{A'} (A':M) \xrightarrow{\text{iso}} A'$. Thus M is projective of rank 1 by 3.4.1: that is (iii) holds.

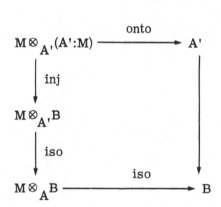

Suppose (ii) holds. Then $(x_\lambda)_{\lambda \in \Lambda}$ bases $M \oplus L$ for some A'-module L. Let $(\chi_\lambda)_{\lambda \in \Lambda}$ be the associated projections: $M \oplus L \to A'$ and let $x_\lambda = m_\lambda + z_\lambda$ for $m_\lambda \in M$ and $z_\lambda \in L$ and $\lambda \in \Lambda$. Now $\frac{s}{1} \in M$ for some $s \in S$, so that $\frac{s}{1} = \sum_{\lambda \in \Lambda} \chi_\lambda(\frac{s}{1}) m_\lambda$ and $\frac{1}{1} = \sum_{\lambda \in \Lambda} \frac{1}{s} \chi_\lambda(\frac{s}{1}) m_\lambda$. Let $x \in M$. Then $\frac{t}{1} x \in A'$ for some $t \in S$ and we have

$$\begin{aligned}
\frac{1}{s} \chi_\lambda(\tfrac{s}{1})\, x &= \frac{1}{st} \chi_\lambda(\tfrac{s}{1}) \frac{t}{1} x \\
&= \frac{1}{st} \chi_\lambda(\tfrac{t}{1} x \tfrac{s}{1}) \\
&= \frac{1}{st} \frac{ts}{1} \chi_\lambda(x) \\
&= \chi_\lambda(x) \\
&\in A'
\end{aligned}$$

for all $\lambda \in \Lambda$. Thus $\frac{1}{s} \chi_\lambda(\frac{s}{1}) \in (A':M)$ for all $\lambda \in \Lambda$ and M is invertible: that is, (i) holds. □

4. Supporting and associated prime ideals

4.1 LENGTHS AND RANKS OF MODULES

An A-module M is called simple iff $M \neq \{0\}$ and its only submodules are $\{0\}$ and M. For example, if $\mathfrak{m}$ is a maximal ideal of A, then $A/\mathfrak{m}$ is a simple A-module. Conversely, every simple A-module is isomorphic to $A/\mathfrak{m}$ for some maximal $\mathfrak{m}$.

An A-module M is called Artinian iff it satisfies the following two equivalent conditions:

(i) every non-empty set of submodules of M has a minimal element;

(ii) every decreasing sequence $N_1 \supseteq N_2 \supseteq \dots$ of submodules of M is eventually constant.

For example, a simple module is both Noetherian and Artinian.

As in 1.3.2, if N is a submodule of M, then M is Artinian iff both N and M/N are Artinian.

A sequence $(N_i)_{0 \leq i \leq n}$ of submodules of an A-module M is called a Jordan-Hölder sequence iff

$$0 = N_0 \subset N_1 \subset \dots \subset N_n = M$$

and N_i / N_{i-1} is simple for $i = 1, \dots, n$. If M has a Jordan-Hölder sequence we say that M is of finite length.

4.1.1 **(Proposition).** (i) *An A-module is of finite length iff it is both Noetherian and Artinian.*

(ii) *Let* M *be an* A-*module and* $(N_i)_{0 \le i \le n}$ and $(L_j)_{0 \le j \le m}$ *be Jordan-Hölder sequences in* M. *Then* $m = n$ *and the* N_i/N_{i-1} *are isomorphic in some order to the* L_j/L_{j-1}.

Proof. (i) Suppose M is Noetherian and Artinian. As long as possible, choose $N_0 = 0$; N_1 to be a minimal element of the set of all submodules $\supset 0$; N_2 to be a minimal element of the set of all submodules $\supset N_1$; and so on. Let N_n be the last possible choice. Then $(N_i)_{0 \le i \le n}$ is a Jordan-Hölder sequence for M.

The converse follows at once from 1.3.2 and its Artinian analogue.

(ii) Using (i) we may suppose that M is the least submodule of M for which the desired result is false. Consider the commutative diagram

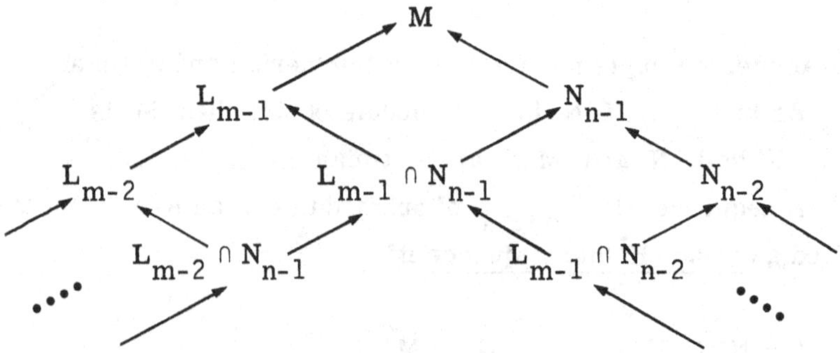

If $L_{m-1} \neq N_{n-1}$, then $L_{m-1} + N_{n-1} = M$ and M/L_{m-1} (resp. M/N_{n-1}) is isomorphic to $N_{n-1}/L_{m-1} \cap N_{n-1}$ (resp. $L_{m-1}/L_{m-1} \cap N_{n-1}$). Moreover the result is true for L_{m-1} and for N_{n-1} and for $L_{m-1} \cap N_{n-1}$: thus it is true for M.

If $L_{m-1} = N_{n-1}$ the result is true for L_{m-1} and therefore for M. (If we work more carefully we do not need to use choice in (ii).) □

4.1.1 (ii) allows us to define the <u>length</u> $\text{long}_A(M)$ of a module M of finite length to be $n \in \omega$ for any Jordan-Hölder sequence $(N_i)_{0 \leq i \leq n}$ in M. If N is a submodule of M, clearly M is of finite length iff both N and M/N are of finite length; and then:

$$\text{long}_A(M) = \text{long}_A(N) + \text{long}_A(M/N) .$$

For example, if A is a field, the A-modules of finite length are the finite-dimensional ones; and their length is their rank.

Let A be an <u>integral</u> ring and k be its field of fractions. Let M be an A-module. We define the <u>torsion submodule</u> $T_A(M)$ of M as follows:

$$T_A(M) = \{m \in M : \lambda m = 0 \text{ for some } \lambda \in A^*\}$$

so that (3.1.2) the sequence $0 \to T_A(M) \to M \to k \otimes_A M$ is exact; and (3.1.3) the functor: $M \longmapsto k \otimes_A M$ (taking A-modules to k-modules) is exact. We say that M is <u>of finite rank</u> iff $k \otimes_A M$ is a k-module of finite type; and we call the rank of $k \otimes_A M$ the <u>rank</u> $r_A(M)$ of M. This definition of rank agrees (where appropriate) with the other ones we have given.

For example M has rank 0 iff M is torsion; and $\mathbf{Q}$ has rank 1 as a $\mathbf{Z}$-module (though $\mathbf{Q}$ is not even of finite type as a $\mathbf{Z}$-module).

If M is an A-module and N is a submodule, the exactness of $M \mapsto k \otimes_A M$ shows that M is of finite rank iff both N and M/N are of finite rank; and that then $r_A(M) = r_A(N) + r_A(M/N)$.

Exercise. If A is an integral ring and not a field, an A-module of finite length is torsion.

4.2 THE SUPPORT OF A MODULE

Let M be an A-module: we define

$$\mathrm{supp}(M) = \{ \mathfrak{p} \in \mathrm{spec}(A) : M_{\mathfrak{p}} \neq 0 \},$$

the support of M.

For example by 3.2.1.1 $\mathrm{supp}(M) = \emptyset$ iff $M = 0$.

4.2.1 (Proposition). Let M be an A-module.

(i) If N is a submodule of M, then $\mathrm{supp}(M) = \mathrm{supp}(N) \cup \mathrm{supp}(M/N)$.

(ii) If $M = \sum_{\lambda \in \Lambda} M_\lambda$ for a family $(M_\lambda)_{\lambda \in \Lambda}$ of submodules, then $\mathrm{supp}(M) = \bigcup_{\lambda \in \Lambda} \mathrm{supp}(M_\lambda)$.

(iii) If M is of finite type, supp(M) is closed in spec(A).

Proof. (i) Since $0 \to N_{\mathfrak{p}} \to M_{\mathfrak{p}} \to (M/N)_{\mathfrak{p}} \to 0$ is exact, we have $M_{\mathfrak{p}} = 0$ iff $N_{\mathfrak{p}} = 0$ and $(M/N)_{\mathfrak{p}} = 0$.

(ii) is similar.

(iii) Let $\mathfrak{a} = \{\lambda \in A: \lambda m = 0$ for all $m \in M\}$, and let $(m_i)_{1 \le i \le n}$ generate M.

Let $\mathfrak{p} \in V(\mathfrak{a})$ and $\mathfrak{p} \notin \text{supp}(M)$. Then for some $s \in A \setminus \mathfrak{p} \subseteq A \setminus \mathfrak{a}$ we have $sm_i = 0$ for $i = 1, \ldots, n$, so that $sM = 0$ and $s \in \mathfrak{a}$, a contradiction.

Conversely let $\mathfrak{p} \in \text{supp}(M)$ and $\lambda \in \mathfrak{a} \setminus \mathfrak{p}$. Then $\lambda m \neq 0$ for some $m \in M$, a contradiction.

Thus $\text{supp}(M) = V(\mathfrak{a})$. □

In particular $\text{supp}(A/\mathfrak{a}) = V(\mathfrak{a})$ for any ideal $\mathfrak{a}$ of A.

4.2.2 **(Proposition).** Let M and N be A-modules of finite type. Then

$$\text{supp}(M \otimes_A N) = \text{supp}(M) \cap \text{supp}(N) .$$

Proof. As in the proof of 3.4.1 $(M \otimes_A N)_{\mathfrak{p}}$ and $M_{\mathfrak{p}} \otimes_{A_{\mathfrak{p}}} N_{\mathfrak{p}}$ are isomorphic $A_{\mathfrak{p}}$-modules. Thus it remains to show that $M \otimes_A N = 0$ implies $M = 0$ or $N = 0$ on the additional hypothesis that A is local. From the natural isomorphism: $(k \otimes_A M) \otimes_k (k \otimes_A N) \to k \otimes_A (M \otimes_A N)$, where k is the residual field of A, we deduce that $k \otimes_A M = 0$ (say), so that $M = 0$ by 3.3.5. □

4.3 PRIME IDEALS ASSOCIATED TO A MODULE

In this section (4.3) A will denote a Noetherian ring. A (*) denotes results where this assumption is not in fact needed.

If M is an A-module and $m \in M$ we define the annihilator ann(m) of m as follows:

$$\text{ann}(m) = \{\lambda \in A: \lambda m = 0\}.$$

We then define ass(M), the set of prime ideals associated to M, as follows:

$$\text{ass}(M) = \{\text{ann}(m) \in \text{spec}(A): m \in M\}.$$

For example, if $\mathfrak{p} \in \text{spec}(A)$ and M is a non-zero submodule of $A/\mathfrak{p}$, then $\text{ass}(M) = \{\mathfrak{p}\}$.

4.3.1 (Proposition). Let M be an A-module.

(*) (i) Every maximal element of the set $X = \{\text{ann}(m): m \in M \text{ and } m \neq 0\}$ belongs to ass(M).

(ii) $M = 0$ iff $\text{ass}(M) = \phi$.

(*) (iii) If N is a submodule of M, then

$$\text{ass}(N) \subseteq \text{ass}(M) \subseteq \text{ass}(N) \cup \text{ass}(M/N).$$

Proof. (i) Let $\text{ann}(m) \in X$ be maximal. Since $m \neq 0$ we have $1 \notin \text{ann}(m)$. Let $x, y \in A$ be such that $xy \in \text{ann}(m)$ but $x \notin \text{ann}(m)$. Then $xm \neq 0$ and $\text{ann}(xm) \in X$. but $\text{ann}(xm) \supseteq \text{ann}(m)$; thus $\text{ann}(xm) = \text{ann}(m)$ and $y \in \text{ann}(xm) = \text{ann}(m)$.

Therefore $\mathrm{ann}(m) \in \mathrm{spec}(A)$ and $\mathrm{ann}(m) \in \mathrm{ass}(M)$.

(ii) Suppose $M \neq 0$. Then $X \neq \phi$ and has a maximal element. Therefore $\mathrm{ass}(M) \neq \phi$ by (i). The converse is immediate.

(iii) Clearly $\mathrm{ass}(N) \subseteq \mathrm{ass}(M)$. Let $\mathfrak{p} = \mathrm{ann}(m) \in \mathrm{ass}(M)$. If $Am \cap N = 0$ we have $\mathfrak{p} = \mathrm{ann}(m + N) \in \mathrm{ass}(M/N)$. If $Am \cap N \neq 0$, the isomorphism: $A/\mathfrak{p} \to Am$ given by $\lambda + \mathfrak{p} \longmapsto \lambda m$ shows that $Am \cap N$ is isomorphic to a non-zero submodule of $A/\mathfrak{p}$, and thus $\{\mathfrak{p}\} = \mathrm{ass}(Am \cap N) \subseteq \mathrm{ass}(N)$. □

(*) **Exercise.** $\mathrm{ass}(\bigoplus_{\lambda \in \Lambda} M_\lambda) = \bigcup_{\lambda \in \Lambda} \mathrm{ass}(M_\lambda)$.

(*) 4.3.2 **(Theorem).** Let M be an A-module and $X \subseteq \mathrm{ass}(M)$. Then there is a submodule N of M such that $\mathrm{ass}(N) = X$ and $\mathrm{ass}(M/N) = \mathrm{ass}(M) \setminus X$.

Proof. Let $Y = \{P \text{ a submodule of } M : \mathrm{ass}(P) \subseteq X\}$. By Zorn's lemma Y has a maximal element N. By 4.3.1 (iii) it suffices now to show that $\mathrm{ass}(M/N) \subseteq \mathrm{ass}(M) \setminus X$. Suppose $\mathfrak{p} \in \mathrm{ass}(M/N)$; then $\mathfrak{p} = \mathrm{ann}(m + N)$ for some $m \notin N$, and

$$\begin{aligned} \mathrm{ass}(Am + N) &\subseteq \mathrm{ass}(N) \cup \mathrm{ass}(Am + N/N) \\ &= \mathrm{ass}(N) \cup \mathrm{ass}(A/\mathfrak{p}) \\ &= \mathrm{ass}(N) \cup \{\mathfrak{p}\}; \end{aligned}$$

but $Am + N \supset N$; so $\mathfrak{p} \notin X$ and $\mathfrak{p} \in \mathrm{ass}(Am + N) \subseteq \mathrm{ass}(M)$. □

<u>4.3.3 **(Proposition).**</u> <u>Let</u> M <u>be an</u> A-<u>module and</u> $\mathfrak{p} \in \mathrm{spec}(A)$. <u>Then</u> $\mathfrak{p} \in \mathrm{supp}(M)$ <u>iff</u> $\mathfrak{p} \supseteq \mathfrak{q}$ <u>for some</u> $\mathfrak{q} \in \mathrm{ass}(M)$.

Proof. Suppose $\mathfrak{p} \supseteq \mathfrak{q}$ for some $\mathfrak{q} = \mathrm{ann}(m) \in \mathrm{ass}(M)$. Then

$$\mathfrak{q}_{\mathfrak{p}} = \mathrm{ann}(\tfrac{m}{1}) \in \mathrm{ass}(M_{\mathfrak{p}})$$

so that by 4.3.1 (ii) $M_{\mathfrak{p}} \neq 0$ and $\mathfrak{p} \in \mathrm{supp}(M)$.

Conversely suppose $\mathfrak{p} \in \mathrm{supp}(M)$; by 4.3.1 (ii) there exists $\mathfrak{P} = \mathrm{ann}(\frac{m}{s}) \in \mathrm{ass}(M_{\mathfrak{p}})$ and $\mathfrak{q} = \{q \in A : \frac{q}{1} \in \mathfrak{P}\} = \mathrm{ann}(m) \in \mathrm{ass}(M)$ and $\mathfrak{p} \supseteq \mathfrak{q}$. □

Thus $\mathrm{ass}(M) \subseteq \mathrm{supp}(M)$ and $\mathrm{ass}(M)$ and $\mathrm{supp}(M)$ have the same minimal elements.

<u>4.3.4 **(Theorem).**</u> <u>Let</u> M <u>be an</u> A-<u>module of finite type.</u>

(i) <u>There are sequences</u> $(N_i)_{0 \le i \le n}$ <u>of submodules of</u> M <u>and</u> $(\mathfrak{p}_i)_{1 \le i \le n}$ <u>in</u> spec(A) <u>such that</u>

(a) $0 = N_0 \subset N_1 \subset \ldots \subset N_n = M$

(b) N_i/N_{i-1} <u>and</u> $A/\mathfrak{p}_i$ <u>are isomorphic for</u> $i = 1, \ldots, n$.

(ii) <u>If</u> $(N_i)_{0 \le i \le n}$ <u>and</u> $(\mathfrak{p}_i)_{1 \le i \le n}$ <u>are such sequences, then:</u>

$$\mathrm{ass}(M) \subseteq \{\mathfrak{p}_1, \ldots, \mathfrak{p}_n\} \subseteq \mathrm{supp}(M) .$$

Proof. (i) Let X be the set of all submodules N of M for which there exist sequences such that (a) and (b) hold with N in place of M. By 1.3.2.2 M is a Noetherian module; and $0 \in X \neq \phi$; so there is a maximal element N of X. Suppose $N \neq M$: then $\mathrm{ass}(M/N) \neq \phi$ by 4.3.1 (ii) and there is a

$\mathfrak{p} = \text{ann}(m + N) \in \text{ass}(M/N)$ for some $m \notin N$. But $Am + N/N$ is isomorphic to $A/\mathfrak{p}$ and $Am + N \in X$, a contradiction. Thus $M \in X$.

(ii) By 4.3.1 (iii) $\text{ass}(M) \subseteq \{\mathfrak{p}_1, \ldots, \mathfrak{p}_n\}$. Also

$$\mathfrak{p}_i \in V(\mathfrak{p}_i) = \text{supp}(A/\mathfrak{p}_i) = \text{supp}(N_i/N_{i-1}) \subseteq \text{supp}(M)$$

by 4.2.1 (i). □

4.3.4.1 (Corollary). ass(M) is finite. □

4.3.4.2 (Corollary). The following conditions are equivalent:

(i) M is of finite length;

(ii) if $\mathfrak{p} \in \text{ass}(M)$, then $\mathfrak{p}$ is a maximal ideal of A;

(iii) if $\mathfrak{p} \in \text{supp}(M)$, then $\mathfrak{p}$ is a maximal ideal of A.

Proof. Suppose (i) holds. Then M has a Jordan-Hölder sequence; and by 4.3.4 (ii) $\text{ass}(M) \subseteq \{\mathfrak{p}_1, \ldots, \mathfrak{p}_n\}$ for some $\mathfrak{p}_i \in \text{spec}(A)$ with $A/\mathfrak{p}_i$ simple, so that $\mathfrak{p}_1, \ldots, \mathfrak{p}_n$ are maximal and (ii) holds.

That (ii) implies (iii) follows at once from 4.3.3.

Suppose (iii) holds. By 4.3.4 (i), (ii) there is a sequence $(N_i)_{0 \leq i \leq n}$ such that $0 = N_0 \subset N_1 \subset \ldots \subset N_n = M$ and there are $\mathfrak{p}_i \in \text{supp}(M)$ such that $A/\mathfrak{p}_i$ and N_i/N_{i-1} are isomorphic for $i = 1, \ldots, n$. But $\mathfrak{p}_i$ is maximal and thus $A/\mathfrak{p}_i$ is simple for $i = 1, \ldots, n$: so $(N_i)_{0 \leq i \leq n}$ is a Jordan-Hölder sequence and (i) holds. □

4.3.4.3 (Corollary). *If* M *is of finite length then* ass(M) = supp(M) *and each has at most* $\text{long}_A(M)$ *elements.* □

4.3.5 (Lemma). *Let* A *be an integral (Noetherian) ring such that if* $\mathfrak{p} \in \text{spec}(A)$, *then* $\mathfrak{p} = 0$ *or* $\mathfrak{p}$ *is maximal.*

(i) *If* M *is a torsion A-module of finite type, then* M *is of finite length.*

(ii) *If* M *is a torsion-free A-module of finite rank, and* $a \in A^*$, *then* M/aM *and* A/Aa *are of finite length and*

$$\text{long}_A(M/aM) \leq r_A(M)\ \text{long}_A(A/Aa)\ .$$

Proof. (i) Since M is torsion, $0 \notin \text{ass}(M)$, so that ass(M) consists of maximal ideals. Thus M is of finite length by 4.3.4.2.

(ii) Let N be a submodule of M *of finite type*. Then $r_A(N) \leq r_A(M)$ and $N \xrightarrow{\text{inj}} k \otimes_A N$, where k is the field of fractions of A; so N has a free submodule L of rank $r_A(N)$; and N/L is torsion and of finite type, thus of finite length by (i). Also by (i) L/a^nL and N/a^nN are of finite length for $n \in \omega$; and we have exact sequences

$$0 \to L/a^nN \cap L \to N/a^nN \to {}^{N/L}\!\big/_{a^n(N/L)} \to 0$$

$$L/a^nL \to L/a^nN \cap L \to 0$$

from which

$$\text{long}_A(N/a^nN) \leq \text{long}_A(L/a^nL) + \text{long}_A(N/L)\ .$$

However N is torsion-free so $N/aN \xrightarrow{\text{iso}} aN/a^2N$ by multiplication by a; so that $\operatorname{long}_A(N/a^nN) = n \operatorname{long}_A(N/aN)$ and

$$\operatorname{long}_A(N/aN) \leq \operatorname{long}_A(L/aL) + \frac{1}{n} \operatorname{long}_A(N/L)$$

for $n \geq 1$. Taking $n > \operatorname{long}_A(N/L)$ we have

$$\operatorname{long}_A(N/aN) \leq \operatorname{long}_A(L/aL) .$$

But L is free and $\operatorname{long}_A(L/aL) = r_A(N) \operatorname{long}_A(A/Aa)$ directly; so that

$$\operatorname{long}_A(N/aN) \leq r_A(M) \operatorname{long}_A(A/Aa) .$$

Suppose now that $(N_i)_{0 \leq i \leq n}$ is a sequence of submodules of M such that

$$aM = N_0 \subset N_1 \subset \dots \subset N_n = M .$$

Let $m_i \in N_i \backslash N_{i-1}$ for $i = 1, \dots, n$. Let $N = \sum_{i=1}^{n} Am_i$ and $M_i = ((\sum_{j=1}^{i} Am_j) + aN)/aN$ for $i = 0, \dots, n$. Then

$$0 = M_0 \subset M_1 \subset \dots \subset M_n = N/aN$$

and $n \leq r_A(M) \operatorname{long}_A(A/Aa)$. Thus M/aM is of finite length $\leq r_A(M) \operatorname{long}_A(A/Aa)$. □

5. Integers

5.1 DEFINITION OF INTEGERS

Let A be a ring and A[X] be the A-algebra of polynomials f(X) over A in an indeterminate X. The A-algebra A[X] has the following universal property: for every A-algebra B and every $x \in B$ there is a unique A-algebra morphism: $A[X] \to B$ such that $X \mapsto x$, namely $a_n X^n + \ldots + a_0 \mapsto a_n x^n + \ldots + a_1 x + a_0 1$. We write f(x) for the image of f(X) under this morphism.

Similar remarks apply to the A-algebra $A[X_1, \ldots, X_n]$ of polynomials in n indeterminates over A.

Exercise. Solve the same universal problems in the category of all rings, not necessarily commutative (but with ones).

A polynomial $f(X) \in A[X]$ is called unitary iff $f(X) = X^n + a_1 X^{n-1} + \ldots + a_n$ for some $n > 0$ and $a_1, \ldots, a_n \in A$.

5.1.1 (Lemma). Let $f(X) \in A[X]$ be unitary.

(i) Let $g(X) \in A[X]$. Then there exist unique $q(X), r(X) \in A[X]$ such that $\deg(r(X)) < \deg(f(X))$ and:

$$g(X) = f(X)q(X) + r(X) .$$

(ii) Let $a \in A$. Then $X - a$ divides $f(X)$ iff $f(a) = 0$.

Proof. Trivial. □

Let B be an A-algebra and $x \in B$. We write $A[x]$ for the least subalgebra of B containing x: so that

$$A[x] = \{f(x) \in B : f(X) \in A[X]\} .$$

Similarly we define $A[x, y]$, etc.

5.1.2 **(Theorem).** Let B be an A-algebra and $x \in B$. The following statements are equivalent:

(i) $f(x) = 0$ for some unitary $f(X) \in A[X]$;

(ii) $A[x]$ is an A-module of finite type;

(iii) there is a subalgebra C of B such that

(a) $x \in C$,

(b) C is an A-module of finite type.

Proof. Suppose (i) holds and let $f(X)$ have degree n. Then $(x^i)_{0 \le i \le n}$ generates the A-module $A[x]$.

Plainly (ii) implies (iii).

Finally suppose (iii) holds. Let $(x_i)_{1 \le i \le n}$ generate the A-module C. Then there exist $a_{ij} \in A$ with

$xx_i = \sum_{j=1}^{n} a_{ij}x_j$ for $i = 1, \ldots, n$. In A define $\delta_{ij} = 0$ if $i \ne j$ and $\delta_{ii} = 1$. In C let $(\beta_{ij})_{1 \le i, j \le n}$ (resp. y) be the adjoint (resp. determinant) of the matrix $(\delta_{ij}x - a_{ij}1)_{1 \le i, j \le n}$. Then

$$\begin{aligned} yx_k &= \sum_{j=1}^{n} \delta_{kj} y x_j \\ &= \sum_{j=1}^{n} \left(\sum_{i=1}^{n} \beta_{ki}(\delta_{ij}x - a_{ij}1)\right)x_j \\ &= \sum_{i=1}^{n} \beta_{ki}\left(\sum_{j=1}^{n} \delta_{ij}xx_j - a_{ij}x_j\right) \\ &= \sum_{i=1}^{n} \beta_{ki}\left(xx_i - \sum_{j=1}^{n} a_{ij}x_j\right) \\ &= 0 \end{aligned}$$

for $k = 1, \ldots, n$. Thus $yz = 0$ for all $z \in C$, and in particular $y = y1 = 0$. But $y = \det((\delta_{ij}x - a_{ij}1)_{1 \le i, j \le n})$ so that $y = f(x)$ for some unitary $f(X) \in A[X]$ and (i) holds. □

5.1.2.1 **(Corollary).** Suppose also that A is Noetherian. We have the further equivalent condition:

(iv) there is a submodule M of B of finite type with $A[x] \subseteq M$.

Proof. 1.3.2.3. □

If $x \in B$ satisfies the conditions of 5.1.2 we say that x is integral over A. If x is integral over A for all $x \in B$, we say that B is an entire A-algebra.

For example, the Gaussian integers comprise an entire **Z**-algebra; but **Q** is not an entire **Z**-algebra. Again, if A and

B are fields, and B is an A-algebra, B is an entire A-algebra iff B is an algebraic extension of A.

Note that if an A-algebra B is an A-module of finite type, it is entire; for example, a finite field extension is algebraic.

5.1.2.2 **(Corollary).** Let $x, y \in B$ be integral over A and $a \in A$. Then $x + y$, xy and ax are integral over A.

Proof. Let $(b_i)_{i \in I}$ (resp. $(c_j)_{j \in J}$) generate A[x] (resp. A[y]) where I and J are finite. Then $(b_i c_j)_{(i,j) \in I \times J}$ generates A[x, y]; and $x + y$, xy, $ax \in A[x, y]$ are integral over A by 5.1.2 (iii). □

By 5.1.2.2 the set A' of all $x \in B$ integral over A is a subalgebra of B. We call A' the integral closure of A in B. If $A' = A1$ we say that A is integrally closed in B. Note that

(i) A' is the largest entire subalgebra of B;

(ii) B is entire iff $B = A'$.

We say that a ring A is integrally closed iff A is integrally closed in tot(A). For example **Z** is integrally closed; and a field is integrally closed.

5.1.2.3 **(Corollary).** Suppose B is an entire A-algebra and C is an entire B-algebra. Then C is an entire A-algebra.

Proof. Let $x \in C$. Then $x^n + b_1 x^{n-1} + \ldots + b_n 1 = 0$ for some $b_1, \ldots, b_n \in B$. As in 5.1.2.2 $A[b_1, \ldots, b_n]$ is an A-module of finite type; and $A[b_1, \ldots, b_n][x]$ is an $A[b_1, \ldots, b_n]$-module of finite type, being generated by $(x^i)_{0 \leq i < n}$; so

$A[b_1, \ldots, b_n][x]$ is an A-module of finite type, and $x \in A[b_1, \ldots, b_n][x]$, so that x is integral over A by 5.1.2 (iii). □

It follows at once that $(A')' = A'$, so that A' is the smallest entire subalgebra of B such that A' is integrally closed in B.

Exercises. (i) If B and C are entire A-algebras, so is $B \otimes_A C$.

(ii) An integral ring with unique factorisation to irreducibles is integrally closed.

5.1.3 (Proposition). Let B be an A-algebra and A' be the integral closure of A in B. Let $S \subseteq A$ be multiplicative. Then $S^{-1}A'$ is the integral closure of $S^{-1}A$ in $S^{-1}B$.

Proof. Let $\frac{b}{s} \in S^{-1}B$ be integral over $S^{-1}A$. Then $\frac{b}{1}$ is integral over $S^{-1}A$ and there exist $a_1, \ldots, a_n \in A$ and $r \in S$ such that

$$\left(\frac{b}{1}\right)^n + \frac{a_1}{r}\left(\frac{b}{1}\right)^{n-1} + \ldots + \frac{a_n}{r}\,\frac{1}{1} = 0\,.$$

Thus for some $t \in S$ we have

$$t(rb^n + a_1 b^{n-1} + \ldots + a_n 1) = 0$$

in B; so that

$$(trb)^n + a_1 t(trb)^{n-1} + \ldots + a_n r^{n-1} t^n 1 = 0$$

and $trb \in A'$. Thus $\frac{b}{s} \in S^{-1}A'$.

Conversely if $b \in A'$ and $s \in S$, we have $b^n + a_1 b^{n-1} + \ldots + a_n 1 = 0$ for some $a_1, \ldots, a_n \in A$; so that

$$\left(\frac{b}{s}\right)^n + \frac{a_1}{s}\left(\frac{b}{s}\right)^{n-1} + \ldots + \frac{a_n 1}{s^n 1} = 0$$

and $\frac{b}{s}$ is integral over $S^{-1}A$. □

5.1.3.1 (Corollary). Let A be an integral ring. Then A is integrally closed iff $A_{\mathfrak{m}}$ is integrally closed for every maximal ideal $\mathfrak{m}$ of A.

Proof. 3.2.1.3. □

5.1.4 (Lemma). Let A be a ring.

(i) Let $f(X) \in A[X]$ be unitary. There is an A-algebra B with $A \xrightarrow{\text{inj}} B$ and $b \in B$ such that $f(b) = 0$ (so that $X - b$ divides $f(X)$ in $B[X]$ by 5.1.1 (ii)).

(ii) Let C be an A-algebra. Let $f(X), g(X) \in C[X]$ be unitary. Then the coefficients of $f(X)g(X)$ are integral over A iff those of both $f(X)$ and $g(X)$ are integral over A.

Proof. (i) Let $f(X)$ generate the ideal $\mathfrak{a}$ of $A[X]$ and let $B = A[X]/\mathfrak{a}$ and $b = X + \mathfrak{a}$. Then $f(b) = f(X) + \mathfrak{a} = 0$ and it remains to show that $A \xrightarrow{\text{inj}} B$. Suppose $a \in A^*$ is such that

$a \in \mathfrak{a}$; then $a = f(X)\,h(X)$ for some non-zero $h(X) \in A[X]$ and a has degree ≥ 1 as a polynomial in $A[X]$, a contradiction.

(ii) Suppose that the coefficients of $f(X)\,g(X)$ are integral over A. Using (i) several times we construct a C-algebra B with $C \xrightarrow{\text{inj}} B$ and such that

$$f(X) = \prod_{i=1}^{n} (X - b_i)$$

$$g(X) = \prod_{j=1}^{m} (X - d_j)$$

for some $b_i, d_j \in B$. Thus the d_i and b_j are integral over A by 5.1.2.3, and the coefficients of $f(X)$ and $g(X)$ (regarded as polynomials in $B[X]$) are integral over A. But $C \xrightarrow{\text{inj}} B$; so the coefficients of $f(X)$ and $g(X)$ (regarded as polynomials over C) are integral over A.

The converse follows at once from 5.1.2.2. □

For example, let A be an integral ring; k be its field of fractions; and K be an algebraic field extension of k. Thus if $x \in K$ there is a unitary $f(X) \in k[X]$ <u>of least degree</u> such that $f(x) = 0$; we call $f(X)$ the <u>minimal</u> polynomial of x over k. If now x is integral over A, then $g(x) = 0$ for some unitary $g(X) \in A[X]$; and by 5.1.1 $g(X) = f(X)\,h(X)$ for some $h(X) \in k[X]$. Thus the coefficients of $f(X)$ are integral over A by 5.1.4. In particular, if A is integrally closed in k, then $x \in K$ is integral over A iff $f(X) \in A[X]$.

5.1.5 **(Proposition).** Let B be an A-algebra and $f(X_1, \ldots, X_n) \in B[X_1, \ldots, X_n]$. Then f is integral over $A[X_1, \ldots, X_n]$ iff its coefficients are integral over A.

Proof. The general case is an immediate induction from the case $n = 1$.

Suppose then that $f(X) \in B[X]$ is integral over $A[X]$: say $H(f(X)) = 0$ where

$$H(Y) = Y^m + F_1(X)Y^{m-1} + \ldots + F_m(X) \in A[X][Y].$$

Let n be strictly greater than the degrees of $f(X)$ and of the $F_i(X)$ and let $f_1(X) = f(X) - X^n$. Thus $-f_1(X)$ is unitary and $H_1(f_1(X)) = 0$ where

$$H_1(Y) = H(Y + X^n) = Y^m + G_1(X)Y^{m-1} + \ldots + G_m(X)$$

so that $G_m(X) = H(X^n)$ is unitary. We have

$$(-f_1(X))(f_1(X)^{m-1} + G_1(X)f_1(X)^{m-2} + \ldots + G_{m-1}(X)) = G_m(X)$$

and the coefficients of $G_m(X)$ lie in A. Thus by 5.1.4 (ii) the coefficients of $f_1(X)$, and therefore of $f(X)$, are integral over A.

The converse is immediate from 5.1.2.2.□

5.1.5.1 **(Corollary).** Let A be an integral ring. Then A is integrally closed iff $A[X_1, \ldots, X_n]$ is integrally closed.

Proof. Suppose first that A is a field. Then A is integrally closed; and $A[X_1, \ldots, X_n]$ has unique factorisation (see last exercise) and is therefore integrally closed.

Suppose then that k is the field of fractions of A and that A is integrally closed. Then $k(X_1, \ldots, X_n)$ is the field of fractions of $A[X_1, \ldots, X_n]$. If $x \in k(X_1, \ldots, X_n)$ is integral over $A[X_1, \ldots, X_n]$, then $x \in k[X_1, \ldots, X_n]$ by the first paragraph, so that $x \in A[X_1, \ldots, X_n]$ by 5.1.5. □

5.2 INTEGERS AND PRIME IDEALS

Let $f:A \to B$ be a ring morphism; $\mathfrak{p} \in \mathrm{spec}(A)$; and $\mathfrak{P} \in \mathrm{spec}(B)$. We say that $\mathfrak{P}$ lies over $\mathfrak{p}$ iff $f^{-1}[\mathfrak{P}] = \mathfrak{p}$: in other words, iff $\mathfrak{p}$ is the image of $\mathfrak{P}$ under the map: $\mathrm{spec}(B) \to \mathrm{spec}(A)$.

In particular, if A is a subring of B, $\mathfrak{P} \in \mathrm{spec}(B)$ lies over $\mathfrak{p} \in \mathrm{spec}(A)$ iff $\mathfrak{P} \cap A = \mathfrak{p}$.

5.2.1 **(Lemma).** Let $f:A \to B$ be a ring morphism and $\mathfrak{p} \in \mathrm{spec}(A)$. Then there is a $\mathfrak{P} \in \mathrm{spec}(B)$ lying over $\mathfrak{p}$ iff $f^{-1}[\mathfrak{p}B] = \mathfrak{p}$.

Proof. Suppose $\mathfrak{P} \in \mathrm{spec}(B)$ and $f^{-1}[\mathfrak{P}] = \mathfrak{p}$. Then $f[\mathfrak{p}] \subseteq \mathfrak{P}$ so that $\mathfrak{p}B \subseteq \mathfrak{P}$. Thus $\mathfrak{p} \subseteq f^{-1}[\mathfrak{p}B] \subseteq f^{-1}[\mathfrak{P}] = \mathfrak{p}$ and $f^{-1}[\mathfrak{p}B] = \mathfrak{p}$.

Conversely suppose $f^{-1}[\mathfrak{p}B] = \mathfrak{p}$ and let $S = f[A \setminus \mathfrak{p}]$. If $s \in A \setminus \mathfrak{p}$ and $f(s) \in \mathfrak{p}B$, then $s \in \mathfrak{p}$, a contradiction. Thus $S \cap \mathfrak{p}B = \emptyset$ and by 1.2.1 there exists $\mathfrak{P} \in \mathrm{spec}(B)$ with $\mathfrak{P} \cap S = \emptyset$ and $\mathfrak{P} \supseteq \mathfrak{p}B$. Plainly $f^{-1}[\mathfrak{P}] = \mathfrak{p}$. □

In particular, if A is a subring of B, and $\mathfrak{p} \in \mathrm{spec}(A)$, there is a $\mathfrak{P} \in \mathrm{spec}(B)$ lying over $\mathfrak{p}$ iff $A \cap \mathfrak{p}B = \mathfrak{p}$; and there is no $\mathfrak{P} \in \mathrm{spec}(B)$ lying over $\mathfrak{p}$ iff $A \cap \mathfrak{p}B \supset \mathfrak{p}$.

5.2.2 **(Proposition).** Let B be an entire A-algebra; $\mathfrak{p} \in \mathrm{spec}(A)$; and $\mathfrak{P} \in \mathrm{spec}(B)$ lie over $\mathfrak{p}$. Then $\mathfrak{p}$ is maximal iff $\mathfrak{P}$ is maximal.

Proof. Since $A/\mathfrak{p} \xrightarrow{\text{inj}} B/\mathfrak{P}$ and $B/\mathfrak{P}$ is an entire $A/\mathfrak{p}$ - algebra, we may suppose that $\mathfrak{p} = 0$ and $\mathfrak{P} = 0$ and $A \xrightarrow{\text{inj}} B$, and must show that A is a field iff B is a field.

Suppose A is a field and let $x \in B^*$. Then $x^n + a_1 x^{n-1} + \ldots + a_n 1 = 0$ for some $a_1, \ldots, a_n \in A$ and least $n \in \omega$. But $a_n \neq 0$, for otherwise $x^{n-1} + a_1 x^{n-2} + \ldots + a_{n-1} 1 = 0$; thus there is an inverse

$$x^{-1} = -a_n^{-1}(x^{n-1} + a_1 x^{n-2} + \ldots + a_{n-1} 1)$$

of x in B and B is a field.

Conversely suppose B is a field. Let $x \in A^*$. Then there is an inverse x^{-1} of x in B; and $x^{-n} + a_1 x^{-n+1} + \ldots + a_n 1 = 0$ for some $a_1, \ldots, a_n \in A$; so $x^{-1} = -(a_1 1 + \ldots + a_n x^{n-1}) \in A$, and A is a field. □

5.2.3 **(Theorem).** (Cohen-Seidenberg). Let B be an entire A-algebra and $A \to B$ be injective. Let $\mathfrak{p} \in \mathrm{spec}(A)$. Then there exists $\mathfrak{P} \in \mathrm{spec}(B)$ lying over $\mathfrak{p}$.

Proof. By 5.1.3 the $A_\mathfrak{p}$ -algebra $B_\mathfrak{p}$ is entire; and by 3.1.3 $A_\mathfrak{p} \xrightarrow{\text{inj}} B_\mathfrak{p}$. Thus $B_\mathfrak{p} \neq 0$. Let $\mathfrak{m}$ be a maximal ideal of $B_\mathfrak{p}$. By 5.2.2 $\mathfrak{m}$ lies over $\mathfrak{p}A_\mathfrak{p}$ in $A_\mathfrak{p}$ and therefore over $\mathfrak{p}$ in A. Let $\mathfrak{m}$ lie over $\mathfrak{P}$ in B. It follows from applying spec to the commutative diagram

$$\begin{array}{ccc} A & \longrightarrow & B \\ \downarrow & & \downarrow \\ A_{\mathfrak{p}} & \longrightarrow & B_{\mathfrak{p}} \end{array}$$

that $\mathfrak{P}$ lies over $\mathfrak{p}$ in A. □

5.2.4 (Lemma). Let B be an entire A-algebra and $A \to B$ be injective. Suppose that B is an integral ring. Let $\mathfrak{P} \in \mathrm{spec}(B)$ lie over $\mathfrak{p} \in \mathrm{spec}(A)$. Then $\mathfrak{P} = 0$ iff $\mathfrak{p} = 0$.

Proof. Let $x \in \mathfrak{P}$ and $x \neq 0$. Then $x^n + a_1 x^{n-1} + \ldots + a_n 1 = 0$ for some $a_1, \ldots, a_n \in A$ and least $n \in \omega$. Thus $n \geq 1$; and $a_n \neq 0$, for otherwise $x^{n-1} + a_1 x^{n-2} + \ldots + a_{n-1} 1 = 0$; and $a_n \in \mathfrak{p}$.

The converse is immediate. □

5.2.4.1 (Corollary). Let B be an entire A-algebra and $\mathfrak{p} \in \mathrm{spec}(A)$. Then the elements of spec(B) which lie over $\mathfrak{p}$ are pairwise incomparable by inclusion.

Proof. Let $\mathfrak{P} \subseteq \mathfrak{Q}$ in spec(B) lie over $\mathfrak{p}$. Then $A/\mathfrak{p} \xrightarrow{\text{inj}} B/\mathfrak{P}$ and $\mathfrak{Q}/\mathfrak{P}$ lies over 0 in $A/\mathfrak{p}$. Thus $\mathfrak{Q} = \mathfrak{P}$. □

5.2.5 (Lemma). Let A be a ring and $\mathfrak{p}_1, \ldots, \mathfrak{p}_n \in \mathrm{spec}(A)$. Let $\mathfrak{q} \in \mathrm{spec}(A)$ with $\mathfrak{q} \not\subseteq \mathfrak{p}_i$ for $i = 1, \ldots, n$. Then there exists $x \in \mathfrak{q}$ such that $x \notin \mathfrak{p}_i$ for all i.

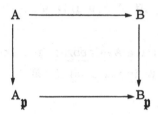

Proof. We use induction on n. Thus we may suppose $n > 1$ and that for each j there exists $x_j \in \mathfrak{q}$ with $x_j \notin \mathfrak{p}_i$ for $i \neq j$. If $x_i \notin \mathfrak{p}_i$ for some i, there is nothing more to prove. Suppose then that $x_i \in \mathfrak{p}_i$ for all i. Let $x = x_1 + \prod_{i=2}^{n} x_i$. Then $x \in \mathfrak{q}$, and $x \notin \mathfrak{p}_1$ since $\prod_{i=2}^{n} x_i \notin \mathfrak{p}_1$ and $x_1 \in \mathfrak{p}_1$; and $x \notin \mathfrak{p}_i$ for $i = 2, \ldots, n$, since $x_1 \notin \mathfrak{p}_i$ and $\prod_{i=2}^{n} x_i \in \mathfrak{p}_i$. □

Let A be an integral ring integrally closed in its field of fractions k. Let K be a finite Galois extension field of k with Galois group Γ. Let B be the integral closure of A in K. If $\sigma \in \Gamma$, then $\sigma[B] \subseteq B$: and we have:

$$B = 1[B] = \sigma[\sigma^{-1}[B]] \subseteq \sigma[B] \subseteq B$$

so $\sigma[B] = B$. Let $\mathfrak{p} \in \operatorname{spec}(A)$ and $\mathfrak{P} \in \operatorname{spec}(B)$ with $A \cap \mathfrak{P} = \mathfrak{p}$; then $\sigma[\mathfrak{P}] \in \operatorname{spec}(B)$ and $A \cap \sigma[\mathfrak{P}] = \mathfrak{p}$ too. Moreover we have a converse result:

<u>5.2.6 **(Theorem).**</u> <u>In the situation just described,</u> Γ <u>acts transitively on the set of all</u> $\mathfrak{P} \in \operatorname{spec}(B)$ <u>lying over</u> $\mathfrak{p}$.

Proof. Let $\mathfrak{P}, \mathfrak{Q} \in \operatorname{spec}(B)$ lie over $\mathfrak{p}$ and suppose $\sigma[\mathfrak{P}] \neq \mathfrak{Q}$ for all $\sigma \in \Gamma$. By 5.2.4.1 $\mathfrak{Q} \not\subseteq \sigma[\mathfrak{P}]$ for all $\sigma \in \Gamma$. Thus by 5.2.5 there exists $x \in \mathfrak{Q}$ such that $x \notin \sigma[\mathfrak{P}]$ for all $\sigma \in \Gamma$. Therefore $\sigma(x) \notin \mathfrak{P}$ for all $\sigma \in \Gamma$ and $\prod_{\sigma \in \Gamma} \sigma(x) = y \notin \mathfrak{P}$.

However $y = (\prod_{1 \neq \sigma \in \Gamma} \sigma(x))\ x \in \mathfrak{Q}$; and $y \in k \cap B = A$; so

$y \in \mathfrak{Q} \cap A = \mathfrak{p} \subseteq \mathfrak{P}$, a contradiction. □

Exercise. What is the analogue of 5. 2. 6 if K is a purely inseparable field extension of k ?

6. Some geometrical results

We say that an A-algebra B is <u>of finite type</u> iff $B = A[x_1, \ldots, x_n]$ for some $x_1, \ldots, x_n \in B$. For example, an entire A-algebra is of finite type iff it is of finite type as an A-module.

Let k be a field and A be a k-algebra. We say that a family $(x_\lambda)_{\lambda \in \Lambda}$ in A is <u>algebraically free</u> iff the unique k-algebra morphism: $k[X_\lambda]_{\lambda \in \Lambda} \to A$ such that $X_\lambda \mapsto x_\lambda$ is injective (so that $k[X_\lambda]_{\lambda \in \Lambda}$ and $k[x_\lambda]_{\lambda \in \Lambda}$ are isomorphic k-algebras).

<u>6.1 **(Theorem)**</u> <u>(Noether's normalisation lemma).</u> <u>Let k be a field and suppose A is a k-algebra of finite type. Then there is an algebraically free family $(x_i)_{1 \le i \le n}$ in A such that A is an entire $k[x_1, \ldots, x_n]$-algebra.</u>

Proof. Let $(y_i)_{1 \le i \le m}$ in A be such that $A = k[y_1, \ldots, y_m]$. We shall use induction on m. If $m = 0$ there is nothing to prove.

Let $m > 0$. Then $k[Y_1, \ldots, Y_m]/\mathfrak{a} \xrightarrow{\text{iso}} A$ for some ideal $\mathfrak{a}$ of $k[Y_1, \ldots, Y_m]$. If $\mathfrak{a} = 0$ we take $(x_i)_{1 \le i \le n} = (y_i)_{1 \le i \le m}$. Suppose therefore that $f(Y_1, \ldots, Y_m) \in \mathfrak{a} \setminus \{0\}$. There is a sequence $(N_i)_{1 \le i \le m-1}$ in ω which increases so rapidly that the leading term of the polynomial

$$f(Y_1 + Y_m^{N_1}, \ldots, Y_{m-1} + Y_m^{N_{m-1}}, Y_m) \in k[Y_1, \ldots, Y_{m-1}][Y_m]$$

belongs to k^*. Let $z_i = y_i - y_m^{N_i}$ for $i = 1, \ldots, m-1$. Then:

$$f(z_1 + y_m^{N_1}, \ldots, z_{m-1} + y_m^{N_{m-1}}, y_m)$$

$$= f(y_1, \ldots, y_m)$$

$$= 0$$

so that y_m is integral over $k[z_1, \ldots, z_{m-1}]$. Therefore $y_1, \ldots, y_{m-1}$ are also integral over $k[z_1, \ldots, z_{m-1}]$ and A is an entire $k[z_1, \ldots, z_{m-1}]$-algebra by 5.1.2.2. However $k[z_1, \ldots, z_{m-1}]$ is an entire $k[x_1, \ldots, x_n]$-algebra for some algebraically free family $(x_i)_{1 \leq i \leq n}$ in $k[z_1, \ldots, z_{m-1}]$ by the inductive hypothesis; thus by 5.1.2.3 A is an entire $k[x_1, \ldots, x_n]$-algebra. □

<u>**6.1.1 (Corollary).**</u> $n \leq m$. □

<u>**6.1.2 (Corollary).**</u> <u>(Weak Nullstellensatz).</u> <u>Let</u> k <u>be a field</u> <u>and</u> $A = k[X_1, \ldots, X_n]$. <u>Let</u> $\mathfrak{p} \in \operatorname{spec}(A)$. <u>Then</u> $\mathfrak{p}$ <u>is maximal</u> <u>iff</u> $A/\mathfrak{p}$ <u>is a</u> k-<u>module of finite type</u>.

Proof. Suppose $\mathfrak{p}$ is maximal. Then by 6.1 $A/\mathfrak{p}$ is an entire $k[y_1, \ldots, y_m]$-algebra for some algebraically free family $(y_i)_{1 \leq i \leq n}$ in $A/\mathfrak{p}$. Thus by 5.2.2 $k[y_1, \ldots, y_m]$ is a field. But $k[y_1, \ldots, y_m]$ is isomorphic to $k[Y_1, \ldots, Y_m]$, so $m = 0$ and $A/\mathfrak{p}$ is an entire k-algebra of finite type. Thus $A/\mathfrak{p}$ is a k-module of finite type, as we remarked above.

Conversely if $A/\mathfrak{p}$ is a k-module of finite type it is a field by 5.2.2. □

APPLICATION TO ALGEBRAIC CLOSURES OF FIELDS

If $(M_\lambda)_{\lambda \in \Lambda}$ is a family of A-modules, we define the tensor product $\bigotimes_{\lambda \in \Lambda} M_\lambda$ as we should expect; but if $(B_\lambda)_{\lambda \in \Lambda}$ is a family of A-algebras, we also define the restricted tensor product $\bigotimes_{\lambda \in \Lambda}^{\text{res}} B_\lambda$ to be the submodule of $\bigotimes_{\lambda \in \Lambda} B_\lambda$ generated by $\bigotimes_{\lambda \in \Lambda} b_\lambda$ for families $(b_\lambda)_{\lambda \in \Lambda} \in \prod_{\lambda \in \Lambda} B_\lambda$ with $b_\lambda = 1$ for all but a finite number of $\lambda \in \Lambda$; and we give $\bigotimes_{\lambda \in \Lambda}^{\text{res}} B_\lambda$ the obvious A-algebra structure. (Clearly $\bigotimes_{\lambda \in \Lambda}^{\text{res}} B_\lambda$ with the natural maps: $B_\lambda \to \bigotimes_{\lambda \in \Lambda}^{\text{res}} B_\lambda$ for each λ is the sum of $(B_\lambda)_{\lambda \in \Lambda}$ in the category of A-algebras.)

Now let k be a field and define

$$A = \bigotimes_{\substack{n \in \omega \\ \mathfrak{m} \text{ maximal}}}^{\text{res}} k[X_1, \ldots, X_n]/\mathfrak{m} .$$

By 6.1.2 and 5.1.2.2 A is an entire k-algebra; and if L is a finite extension field of k, there is a k-algebra morphism: $L \to A$.

By flatness A is non-zero. Let $\mathfrak{P}$ be any prime ideal of A. Then by 5.2.2 $k \to A/\mathfrak{P}$ is an algebraic closure of k.

(Compare any other construction of algebraic closures.)

6.1.3 (Corollary) (Hilbert's Nullstellensatz). Let k be a field; $A = k[X_1, \ldots, X_n]$; $\mathfrak{a}$ be an ideal of A; and $x \in A$. Suppose that for every finite extension field L of k and every k-algebra morphism $\phi: A \to L$ with $\mathfrak{a} \subseteq \operatorname{Ker} \phi$ we have $\phi(x) = 0$. Then $x^n \in \mathfrak{a}$ for some $n \in \omega$.

Proof. We may suppose $x \neq 0$.

In the A-algebra $B = k[X_1, \ldots, X_n, X_{n+1}]$ let $\mathfrak{b} = B\mathfrak{a} + B(1 - xX_{n+1})$. Suppose $\mathfrak{b} \subset B$. Then $\mathfrak{b} \subseteq \mathfrak{m}$ for some maximal ideal $\mathfrak{m}$ of B; and $B/\mathfrak{m}$ is a finite extension field of k by 6.1.2. Let $\phi: A \to B \to B/\mathfrak{m}$ be the natural map. Then $\mathfrak{a} \subseteq \operatorname{Ker} \phi$ so $\phi(x) = 0$ and $\phi(1) = \phi(1 - xX_{n+1}) + \phi(x)\phi(X_{n+1}) = 0$, a contradiction.

Thus $\mathfrak{b} = B$ and

$$1 = \sum_{i=1}^{r} f_i(X_1, \ldots, X_n)g_i(X_1, \ldots, X_{n+1}) + (1 - xX_{n+1})g(X_1, \ldots, X_{n+1})$$

for some $f_i \in \mathfrak{a}$ and $g, g_i \in B$. Now work in $k(X_1, \ldots, X_{n+1})$ and put $\frac{1}{x}$ for X_{n+1} to obtain

$$1 = \sum_{i=1}^{r} f_i(X_1, \ldots, X_n)g_i(X_1, \ldots, X_n, \tfrac{1}{x})$$

so that $x^n \in \mathfrak{a}$ for some $n \in \omega$. □

7. Valuation rings

7.1 ORDERED GROUPS

A *partially ordered abelian group* $(G, +, \leq)$ is an abelian group $(G, +)$ with a partial order $\leq$ on G such that:

$$x \leq y \Rightarrow x + z \leq y + z$$

for all $x, y, z \in G$.

We say that G is *replete* iff for all $n \in \omega$ with $n > 0$ we have:

$$x \in G \text{ and } nx > 0 \Rightarrow x > 0 .$$

For example, if G is totally ordered, it is replete.

7.1.1 (Lemma). *Let G be a replete partially ordered torsion-free abelian group; let H be a totally ordered subgroup; and G/H be torsion. Then G is totally ordered.*

Proof. Let $x \in G$ with $x \neq 0$. Then $nx \in H$ and $nx \neq 0$ for some $n > 0$. Thus (say) $nx > 0$ and $x > 0$. □

Now let A be an integral ring and k be its field of fractions. We make $k^*/\mathfrak{U}(A)$ into a partially ordered abelian group under multiplication by:

$$x\,\mathfrak{u}(A) \le y\,\mathfrak{u}(A) \text{ iff } xy^{-1} \in A .$$

7.1.2 (Lemma). If A is integrally closed in k, then $k^*/\mathfrak{u}(A)$ is replete and torsion-free.

Proof. If $x^n \in \mathfrak{u}(A)$, then $x \in \mathfrak{u}(A)$; if $x^n \in A\setminus\mathfrak{u}(A)$, then $x \in A\setminus\mathfrak{u}(A)$. □

7.2 VALUATION RINGS

7.2.1 (Proposition). Let A be an integral ring and k be its field of fractions. The following statements are equivalent:

(i) $k^*/\mathfrak{u}(A)$ is totally ordered;

(ii) $x \in k\setminus A \Rightarrow x^{-1} \in A$;

(iii) the ideals of A are totally ordered by inclusion.

Proof. Immediate. □

We call such a ring A a valuation ring (in k, if we wish to specify its field of fractions). For example, $\mathbf{Q}$ and $\mathbf{Z}_{p\mathbf{Z}}$ for p prime are valuation rings in $\mathbf{Q}$. Note that (ii) already implies that k is the field of fractions of A. By (ii) A is integrally closed in k; and by (iii) A is a local ring.

We call $k^*/\mathfrak{u}(A)$ the value group of A and write $|\ |$ for the natural map: $k^* \to k^*/\mathfrak{u}(A)$ and define $|0| = 0$. Then it is plain that $|x + y| \le \sup(|x|, |y|)$ for all $x, y \in k$, so that if $|x| < |y|$, then $|x + y| = |y|$. It follows that A is Bezout: for if x, y $\in$ A, then (say) $|x| \leq |y|$ and $Ax + Ay = Ay$.

Exercise. A local Bezout ring is a valuation ring.

Let K be a field extension of k and A (resp. B) be a valuation ring in k (resp. K). We say that B extends A iff $B \cap k = A$. (In any case it follows from 7. 2. 1 (ii) that $B \cap k$ is a valuation ring in k.) If B extends A, we find

$$A \cap \mathfrak{m}(B) = \mathfrak{m}(A)$$

$$k^* \cap \mathfrak{u}(B) = \mathfrak{u}(A)$$

and have injections:

$$A/\mathfrak{m}(A) \to B/\mathfrak{m}(B)$$

$$k^*/\mathfrak{u}(A) \to K^*/\mathfrak{u}(B)$$

preserving the field and ordered group structures. The dimension of $\kappa(B)$ as a $\kappa(A)$-module is called the residual degree $f(B|A)$ of the extension; the index of $|k^*|$ as a subgroup of $|K^*|$ is called the ramification $e(B|A)$ of the extension.

7. 2. 2 **(Proposition).** If K is a finite (resp. algebraic) extension field of k, so is $\kappa(B)$ of $\kappa(A)$; and $|K^*|/|k^*|$ is a finite (resp. torsion) group.

Proof. Suppose K is a finite extension of k. Note that $K^*/k^*\mathfrak{u}(B)$ and $|K^*|/|k^*|$ are isomorphic. Let $x_1, \ldots, x_r \in K^*$ be in distinct cosets of $k^*\mathfrak{u}(B)$. Then for any $\lambda_1, \ldots, \lambda_r \in k^*$

the values $|\lambda_i x_i|$ are all different: so $\left|\sum_{i=1}^{r} \lambda_i x_i\right| = \sup_{i=1}^{r} |\lambda_i x_i| \neq 0$ and $\sum_{i=1}^{r} \lambda_i x_i \neq 0$. Thus $r \leq |K:k|$.

The other cases are similar. □

A valuation ring which is a field is called <u>trivial.</u> From 7.2.2, if K is an algebraic extension field of k and B in K extends A in k, then B is trivial iff A is trivial.

If A is a valuation ring in k and $k^*/\mathfrak{u}(A)$ is infinite cyclic we call A <u>discrete.</u> In this case there is a unique decreasing isomorphism: $k^*/\mathfrak{u}(A) \to \mathbf{Z}$, and we write ord for the induced map: $k^* \to \mathbf{Z}$, defining also $\mathrm{ord}(0) = \infty$. Thus if $\mathfrak{a}$ is a non-zero ideal of A and $\mathrm{ord}(\pi) = 1$ and $n = \inf_{x \in \mathfrak{a}} \mathrm{ord}(x)$, then $\mathfrak{a} = \pi^n A$. Therefore A is a principal ring.

For example $\mathbf{Z}_{p\mathbf{Z}}$ is a discrete valuation ring in $\mathbf{Q}$ and $\mathrm{ord}(\frac{a}{b} p^n) = n$ if $a, b \in \mathbf{Z}$ and $(a, p) = (b, p) = 1$. Again, if $K = k(X_1, \ldots, X_n)$ for a field k, we define the <u>total degree valuation ring</u> A in K to consist of $\frac{f}{g}$ for $f, g \in k[X_1, \ldots, X_n]$ with $g \neq 0$ and $\deg(f) \leq \deg(g)$. Clearly A is a discrete valuation ring.

From 7.2.2, if K is a finite extension of k and B in K extends A in k, then B is discrete iff A is discrete: for

$$k^*/\mathfrak{u}(A) \xrightarrow{\text{inj}} K^*/\mathfrak{u}(B) \xrightarrow[\text{inj}]{x \mapsto x^{e(B|A)}} k^*/\mathfrak{u}(A).$$

Exercise. Let k be a finite field extension of $\mathbf{Q}$. Show that every non-trivial valuation ring in k is discrete, and that its residual field is finite.

7.2.3 **(Theorem).** Let A be an integral ring; k be its field of fractions; and $A \subset k$. The following conditions are equivalent:

(i) A is a discrete valuation ring;

(ii) A is a Noetherian local ring and $\mathfrak{m}(A)$ is principal;

(iii) A is a Noetherian valuation ring;

(iv) A is a Noetherian; A is integrally closed in k; and $\text{card}(\text{spec}(A)) = 2$.

Proof. We have seen that (i) implies (iii). Since a valuation ring is Bezout, (iii) implies (ii). If (i) holds and $\mathfrak{p} \in \text{spec}(A)$ with $\mathfrak{p} \neq 0$, then $\mathfrak{p} = \pi^n A$ for some $n > 0$ with $\pi \in A$ with $\text{ord}(\pi) = 1$; and if $n > 1$, we have $\pi\pi^{n-1} \in \mathfrak{p}$ but $\pi, \pi^{n-1} \notin \mathfrak{p}$; so that $n = 1$ and (iv) holds.

Suppose (ii) holds, with $\mathfrak{m}(A) = \pi A$ say. Let $y \in \bigcap_{n \in \omega} \pi^n A$. Then for all $n \in \omega$ $y = \pi^n x_n$ for some $x_n \in A$. If $y \neq 0$ we have $x_n = \pi x_{n+1} \in A^*$ for all $n \in \omega$, so that $Ax_0 \subset Ax_1 \subset Ax_2 \subset \dots$, a contradiction. Thus $\bigcap_{n \in \omega} \pi^n A = 0$. Let $x \in A^*$. Then $x \notin \pi^{n+1} A$ for some least $n \in \omega$ and $\pi^{-n} x \in \mathfrak{u}(A)$. Thus $\pi\,\mathfrak{u}(A)$ generates $k^*/\mathfrak{u}(A)$; and $\pi\,\mathfrak{u}(A) < 1$; so that $k^*/\mathfrak{u}(A)$ is totally ordered and infinite cyclic; that is, (i) holds.

Finally, suppose (iv) holds. Then A is local and $\mathfrak{m}(A) = \mathfrak{m} \neq 0$. We have $(A:\mathfrak{m}) \subset y^{-1}A$ for any $y \in \mathfrak{m} \cap A^*$ (where $(A:\mathfrak{m})$ is defined as in 3.4 with $B = k$). Thus by 1.3.2.3 $(A:\mathfrak{m})$ is an A-module of finite type.

Let $x \in \mathfrak{m}$ and $x \neq 0$. Suppose that $\mathfrak{n}$ is a maximal ideal of A_x. Then $x \notin \mathfrak{n}$ so $\mathfrak{n} \cap A = 0$. Let $\frac{y}{x^n} \in \mathfrak{n}$; then

$y \in \mathfrak{n} \cap A = 0$ and $\frac{y}{x^n} = 0$. Thus $\mathfrak{n} = 0$ and $A_x = k$. Now let $z \in \mathfrak{m} \cap A^*$. Then $\frac{x^n}{z} \in A$ for some $n \in \omega$, and $x^n \in zA$. Thus (since $\mathfrak{m}$ is an A-module of finite type, and using the multinomial theorem) $\mathfrak{m}^N \subseteq zA$ for some least $N \in \omega$. Let $y \in \mathfrak{m}^{N-1} \backslash zA$. Then $\frac{y}{z} \in (A: \mathfrak{m}) \backslash A$ and $(A: \mathfrak{m}) \supset A$.

We have $\mathfrak{m} \subseteq \mathfrak{m}(A: \mathfrak{m}) \subseteq A$, so that $\mathfrak{m}(A: \mathfrak{m})$ is $\mathfrak{m}$ or A. Suppose $\mathfrak{m}(A: \mathfrak{m}) = \mathfrak{m}$ and let $x \in (A: \mathfrak{m})$. Then $x\mathfrak{m} \subseteq \mathfrak{m}$ and $x^n \in (A: \mathfrak{m})$ for all $n \in \omega$. Let $\mathfrak{a}_i = A + Ax + \ldots + Ax^i$ for $i \in \omega$. Then $\mathfrak{a}_0 \subseteq \mathfrak{a}_1 \subseteq \ldots \subseteq (A: \mathfrak{m})$ so that by 1. 3. 2. 3 $\mathfrak{a}_n = \mathfrak{a}_{n-1}$ for some $n \in \omega$. Thus $x^n = b_0 + b_1 x + \ldots + b_{n-1}x^{n-1}$ and $x \in A$. That is, $(A: \mathfrak{m}) \subseteq A$, a contradiction. Thus $\mathfrak{m}(A: \mathfrak{m}) = A$: that is, $\mathfrak{m}$ is invertible, so that $\mathfrak{m} = \pi A$ for some $\pi \in A$ by 3. 4. 3 and 3. 3. 6. That is, (ii) holds. □

7. 3 EXTENSION THEOREMS

Among all pairs $(A, \mathfrak{p})$ of subrings A of a field k and $\mathfrak{p} \in \operatorname{spec}(A)$ let us define the partial order $\leq$ of <u>domination:</u>

$$(A, \mathfrak{p}) \leq (A', \mathfrak{p}') \text{ iff } A \subseteq A' \text{ and } A \cap \mathfrak{p}' = \mathfrak{p}.$$

Thus the injection: $A/\mathfrak{p} \to A'/\mathfrak{p}'$ induces an injection: $L \to L'$ of their fields of fractions and we define as follows the partial order $\leq^*$ of <u>strong domination:</u>

$(A, \mathfrak{p}) \leq^* (A', \mathfrak{p}')$ iff:

(i) $(A, \mathfrak{p}) \leq (A', \mathfrak{p}')$

(ii) L' is an algebraic extension field of L.

In particular, if A and B are local rings in k, we define $A \leq B$ (resp. $A \leq^* B$) iff $(A, \mathfrak{m}(A)) \leq (B, \mathfrak{m}(B))$ (resp. $(A, \mathfrak{m}(A)) \leq^* (B, \mathfrak{m}(B))$).

7.3.1 (Lemma). *Let A and B be local subrings of a field k and $A \leq B$. Suppose that A is a valuation ring in k. Then $A = B$.*

Proof. Immediate. □

7.3.2 (Lemma). *Suppose $(A, \mathfrak{p}) < (A', \mathfrak{p}')$ in k. Then $(A, \mathfrak{p}) <^* (A'', \mathfrak{p}'')$ for some $(A'', \mathfrak{p}'')$ in k.*

Proof. Let $f: A' \to L'$ be the natural map.

If L' is an algebraic extension of L there is nothing to prove. Thus we may suppose that there is a $t \in A'$ such that $f(t) \in L'$ is transcendental over L. Let $g: L[f(t)] \to L$ be the L-algebra morphism: $f(t) \longmapsto 0$. Then we take $A'' = A[t]$ and $\mathfrak{p}'' = \mathrm{Ker}(g \circ f|_{A[t]})$, so that $(A, \mathfrak{p}) < (A'', \mathfrak{p}'')$ and $L'' = L$. □

Thus the maximal pairs in k with respect to $\leq$ and to $\leq^*$ are the same.

7.3.3 (Theorem) (Chevalley). *Let $(A, \mathfrak{p})$ be a pair in k. Then there is a valuation ring B in k with $(A, \mathfrak{p}) \leq^* (B, \mathfrak{m}(B))$.*

Proof. By Zorn's lemma, among all pairs $\geq^* (A, \mathfrak{p})$, ordered by $\leq^*$, there is a maximal one, $(B, \mathfrak{q})$ say. By 5. 2. 3, B is integrally closed in k. Since $(B_{\mathfrak{q}}, \mathfrak{q} B_{\mathfrak{q}}) \geq^* (B, \mathfrak{q})$, B is local and $\mathfrak{m}(B) = \mathfrak{q}$.

It remains to show that B is a valuation ring in k. Suppose $x \in k \backslash B$. By 7. 3. 2 and 5. 2. 1 $(\mathfrak{q} B[x]) \cap B \supset \mathfrak{q}$ so that $1 \in \mathfrak{q} B[x]$. Thus $1 = q_0 + q_1 x + \ldots + q_n x^n$ for some $q_i \in \mathfrak{q}$; and:

$$(x^{-1})^n + (q_0 - 1)^{-1} q_1 (x^{-1})^{n-1} + \ldots + (q_0 - 1)^{-1} q_n = 0 .$$

That is, x^{-1} is integral over B: so that $x^{-1} \in B$. □

Thus the valuation rings in k are the local rings which are maximal for the orders both of domination and of strong domination.

7. 3. 3. 1 **(Corollary).** *If K is a field extension of k and A is a valuation ring in k, there is a valuation ring B in K extending A.*

Proof. There is a valuation ring B in K with $(B, \mathfrak{m}(B)) \geq^* (A, \mathfrak{m}(A))$. Thus $B \cap k = A$ by 7. 2. 1 (ii). □

7. 3. 3. 2 **(Corollary).** *Let A be an integrally closed subring in a field k. Then $A = \cap \{B; B \supseteq A$ and B is a valuation ring in $k\}$.*

Proof. Suppose $x \in k \backslash A$. Then $x \notin A[x^{-1}]$ so $x^{-1} \notin \mathfrak{u}(A[x^{-1}])$ and $x^{-1} \notin \mathfrak{p}$ for some $\mathfrak{p} \in \operatorname{spec}(A[x^{-1}])$. There is a valuation ring B in k with $(B, \mathfrak{m}(B)) \geq (A[x^{-1}], \mathfrak{p})$; thus $x^{-1} \in \mathfrak{m}(B)$ and $x \notin B$. □

From 7.3.3.2 we see that the integral closure of a subring A in a field k is the intersection of all the valuation rings B in k such that $B \supseteq A$.

7.3.3.3 **(Corollary).** If in 7.3.3.2 A is also local, we may further restrict B to dominate A.

Proof. We let $x \in k \backslash A$ and $x^{-1} \in \mathfrak{m}$ for some maximal ideal $\mathfrak{m}$ of A. The composed map: $A \to A[x^{-1}] \to A[x^{-1}]/\mathfrak{m}$ is onto; so its kernel, namely $A \cap \mathfrak{m}$, is maximal in A. Thus $A \cap \mathfrak{m} = \mathfrak{m}(A)$ and B taken as before dominates A. □

7.3.4a **(Lemma).** Let K be an algebraic extension field of k and A be a valuation ring in k. Let B be the integral closure of A in K, C a valuation ring extending A, and $\mathfrak{P} = \mathfrak{m}(C) \cap B$. Then $C = B_{\mathfrak{P}}$.

Proof. As in 7.3.3.2 we see $B_{\mathfrak{P}} \subseteq C$. Conversely suppose $x \in C$; then for some $a_0, \ldots, a_n \in A$, not all zero:

$$a_n x^n + \ldots + a_0 = 0.$$

Choose s as the largest integer such that $0 \leq s \leq n$ and $|a_s| = \max_{0 \leq i \leq n} |a_i|$; then putting $b_i = a_i/a_s$ and dividing through by $a_s x^s$ we have

$$(b_n x^{n-s} + \ldots + b_{s+1} + 1) + \frac{1}{x}(b_{s-1} + \ldots + b_0 \frac{1}{x^{s-1}}) = 0$$

with $b_n, \ldots, b_{s+1} \in \mathfrak{m}(A) \subseteq \mathfrak{m}(C)$ (since $|a_s| > |a_i|$ if

$s \leq i \leq n$) and $b_{s-1}, \ldots, b_0 \in A$ (since $|a_s| \geq |a_i|$). Write $b_n x^{n-s} + \ldots + b_{s+1} + 1 = y$, $b_{s-1} + \ldots + b_0(1/x^{s-1}) = z$, so that $y + \frac{z}{x} = 0$. We show $z \in B$, $y \in B \setminus \mathfrak{P}$. Suppose D is a valuation ring in K such that $D \supseteq B$. If $x \in D$, then $y \in D$ and so $z = -xy \in D$; if $\frac{1}{x} \in D$, then $z \in D$ and so $y = -z . \frac{1}{x} \in D$. Thus in any case both $y, z \in D$. Hence by 7. 3. 3. 2 $y, z \in B$. Further, as $b_n, \ldots, b_{s+1} \in \mathfrak{m}(C)$, $y \notin \mathfrak{m}(C)$, so $y \notin \mathfrak{P}$. □

7. 3. 4 (Theorem). *Let K be an algebraic extension field of k, and A be a valuation ring in k. Let B be the integral closure of A in K. Then the valuation rings in K which extend A are precisely the rings $B_{\mathfrak{P}}$ as $\mathfrak{P}$ runs through the maximal ideals of B.*

Proof. If C is a valuation ring in K extending A, we have by 5. 2. 2 that $\mathfrak{P} = B \cap \mathfrak{m}(C)$ is maximal, and by 7. 3. 4a $C = B_{\mathfrak{P}}$.

Conversely for $\mathfrak{P}$ maximal in B by 7. 3. 3, the local ring $B_{\mathfrak{P}}$ is dominated by a valuation ring C, and by 7. 3. 4a $C = B_{\mathfrak{P}'}$, for $\mathfrak{P}' = \mathfrak{m}(C) \cap B \supseteq \mathfrak{P}$, so that $\mathfrak{P}' = \mathfrak{P}$. □

7. 3. 4. 1 (Corollary). *Suppose in addition that K is a finite Galois extension of k with Galois group Γ. Then Γ acts transitively on the valuation rings in K which extend A.*

Proof. Immediate from 5. 2. 6. □

7. 3. 4. 2 (Corollary). *Suppose instead that K is a purely inseparable extension of k. Then there is one and only one valuation ring in K which extends A, namely B.*

Proof. By 7.3.1 it is enough to show that B is a valuation ring in K. Let p be the characteristic of k and define $B' = \{x \in K: x^{p^n} \in A \text{ for some } n \in \omega\} \subseteq B$. By (ii) of 7.2.1 we see that B' is a valuation ring in K; and if $x^{p^n} \in A$ and $x \in k \backslash A$, then $x^{-1} \in A$ and $x = x^{p^n}(x^{-1})^{p^n - 1} \in A$; so $B' \cap k = A$. But B' is integrally closed in K: so B' = B. □

7.3.4.3 **(Corollary).** Suppose that K is a finite extension of k. Then there are only finitely many valuation rings in K which extend A.

Proof. Fit 7.3.4.1 and 7.3.4.2 together with standard field theory. □

7.4 AN APPLICATION

7.4.1 **(Lemma).** Let A be a principal ring and M be a free A-module. Let L be a submodule of M. Then L is free.

Proof. Let $(m_\beta)_{\beta \in \alpha}$ base M, where α is an ordinal, and let $(\chi_\beta : M \to A)_{\beta \in \alpha}$ be the associated projections. Let $M_\beta = \sum_{\gamma \subseteq \beta} Am_\gamma$ and $L_\beta = L \cap M_\beta$ for all $\beta \in \alpha$, so that $L = \bigcup_{\beta \in \alpha} L_\beta$. For each $\beta \in \alpha$ choose $x_\beta \in L_\beta$ such that

(i) if $\chi_\beta[L_\beta] = 0$, then $x_\beta = 0$;

(ii) if $\chi_\beta[L_\beta] \neq 0$, then $A\chi_\beta(x_\beta) = \chi_\beta[L_\beta]$.

Let $N_\beta = \sum_{\gamma \subseteq \beta} Ax_\gamma \subseteq L_\beta \subseteq M_\beta$. We shall show by transfinite induction that $N_\beta = L_\beta$ for all $\beta \in \alpha$. Suppose then that $N_\gamma = L_\gamma$ for all $\gamma \in \beta$, and let $x \in L_\beta$. If $\chi_\beta[L_\beta] = 0$, then $x \in L_\gamma = N_\gamma \subseteq N_\beta$ for some $\gamma \in \beta$. If $\chi_\beta[L_\beta] \neq 0$, then $\chi_\beta(x) = y\chi_\beta(x_\beta)$ for some $y \in A$, so that $\chi_\beta(x - yx_\beta) = 0$ and $x - yx_\beta \in L_\gamma = N_\gamma \subseteq N_\beta$ for some $\gamma \in \beta$. Thus $x \in N_\beta$; and $N_\beta = L_\beta$ for all $\beta \in \alpha$.

Therefore $L = \sum_{x_\beta \neq 0} Ax_\beta$.

Suppose $\sum_{i=1}^{r} \xi_{\beta_i} x_{\beta_i} = 0$ for some $\xi_{\beta_i} \in A^*$, non-zero x_{β_i}, and $\beta_1 \in \ldots \in \beta_r$. Then

$$0 = \sum_{i=1}^{r} \xi_{\beta_i} \chi_{\beta_r}(x_{\beta_i}) = \xi_{\beta_r} \chi_{\beta_r}(x_{\beta_r}) \neq 0$$

a contradiction.

Thus $(x_\beta)_{x_\beta \neq 0}$ bases L. □

(We used choice when we supposed that α was an ordinal.)

7.4.1.1 **(Corollary).** If M is of finite type, so is L. □

7.4.2 **(Theorem)** (Samuel). Let k be a field and A be an integral k-algebra of finite type. Let K be the integral closure of k in A (so that K is a field). Then under multiplication $\mathfrak{u}(A)/K^*$ is a free $\mathbf{Z}$-module of finite type.

Proof. By 6.1 A is an entire $K[x_1, \ldots, x_n]$-algebra for some algebraically free family $(x_i)_{1 \leq i \leq n}$ in A. Let L be the field of fractions of A, so that L is a finite extension field of $K(x_1, \ldots, x_n)$. Let M be a splitting field for L over $K(x_1, \ldots, x_n)$ with Galois group Γ. Let B be the total degree valuation ring in $K(x_1, \ldots, x_n)$ and let $C_1, \ldots, C_m$ be its extensions to M. These are finitely many and Γ is transitive on them by 7.3.4.1, 2, 3; and they are discrete by 7.2.2. Let us define ord and ord_i to correspond.

Map $\mathfrak{u}(A) \to \mathbf{Z}^m$ by

$$u \mapsto (\mathrm{ord}_1(u), \ldots, \mathrm{ord}_m(u)) .$$

Plainly $K^* \to \{(0, \ldots, 0)\}$. Conversely let $x \in \mathfrak{u}(A)$ and $\mathrm{ord}_i(x) = 0$ for all i. Let $y_1, \ldots, y_r$ be the distinct members of $\{\sigma(x): \sigma \in \Gamma\}$, so that

$$g(X) = \prod_{i=1}^{r} (X - y_i) = X^r + a_1 X^{r-1} + \ldots + a_r$$

is the minimal polynomial of x over $K(x_1, \ldots, x_n)$. (A little care is needed to see this, especially in characteristic p.) Thus $\mathrm{ord}(a_i) \geq 0$ for all i, for $\mathrm{ord}_i(y_j) \geq 0$ for all i, j. However $a_i \in K[x_1, \ldots, x_n]$ for each i by 5.1.5 and the remarks after 5.1.4: thus $a_i \in K$ for all i and $x \in K^*$.

Therefore $\mathfrak{u}(A)/K^* \to \mathbf{Z}^m$ is a monomorphism of $\mathbf{Z}$-modules and $\mathfrak{u}(A)/K^*$ is a free $\mathbf{Z}$-module of finite type by 7.4.1 and 7.4.1.1.□

Bibliography.

Schilling, Valuations (dull).

Ax and Kochen, Amer. J. Math. 87 (1965), 605-730 (beautiful but undigested).

EXAMPLES ON GENERAL VALUATIONS

These examples get harder and harder. Numbers 3 to 5 should be done consecutively, as should numbers 6 to 11.

1. Find all valuation rings of the field **Q** of rational numbers. [Hint: consider the intersections of their maximal ideals with **Z**.]

2. Considering $|x - y|$ as the distance between x and y, show that every triangle is isosceles, and that every point inside a circle is a centre for the circle, with the same radius. [Hint: if $|x| < |y|$, then $|x + y| = |y|$.]

3. If G is a group and F is a field, define F[G] to be the set of all maps $f: G \to F$ such that $f^{-1}F^*$ is finite. On F[G] make the following definitions:

$$(f + g)(\sigma) = f(\sigma) + g(\sigma)$$

$$(\lambda f)(\sigma) = \lambda f(\sigma)$$

$$(fg)(\sigma) = \sum_{\tau \in G} f(\tau)g(\tau^{-1}\sigma),$$

for f, $g \in F[G]$, $\lambda \in F$ and $\sigma \in G$.

Show that F[G] is an F-algebra with a 1. [Hint: it might be easier to inject $G \to F[G]$ as follows: $\sigma \longmapsto f_\sigma$, where $f_\sigma(\sigma) = 1$ and $f_\sigma(\tau) = 0$ if $\tau \neq \sigma$; then an element of F[G] is a finite linear combination of elements of G, with natural definitions of operations.]

(F[G] is called the <u>group algebra</u> of G over F.)

4. If G is a totally ordered abelian group, $F[G]$ is an integral domain.

5. If G is a totally ordered abelian group and F is a field, there is a field k and a valuation $|\ |$ on k, for which $|k^*| = G$, and whose residue class field is F. [Hint: let k be the quotient field of $F[G]$.]

6. Let K/k be a field extension. We say that a family $(\alpha_\lambda)_{\lambda \in \Lambda}$ of elements of K is <u>algebraically independent over</u> k if and only if the (larger) family of all finite products (including the empty product) of elements of $(\alpha_\lambda)_{\lambda \in \Lambda}$ is linearly independent over k. In other words, if and only if the α_λ satisfy no polynomial equation with coefficients in k.

Use Zorn's lemma to show that there exist maximal algebraically independent families in K. Such a family we call a <u>transcendence base</u> of K/k.

7. If L is obtained from k by adjoining a transcendence base of K/k, show that K/L is algebraic.

8. We write $[K{:}k]$ for the number of elements in a transcendence base (it may not be finite: we make no distinction among infinite cardinals). Show that $[K{:}k]$ is well-defined (i. e. independent of the choice of transcendence base). [Hint: use 7.]

9. For this definition, we need not restrict K to be a field. All it need be is a commutative k-algebra with no divisors of zero. Elucidate this statement, and show that if in this wider situation, $\mathfrak{p}$ is a prime ideal of K, then

$$[K{:}k] \geq [\mathfrak{p}{:}k] + [K/\mathfrak{p}{:}k].$$

10. If G is an abelian group, then the set $\{g_1, \ldots, g_r\} \subseteq G$ is called <u>independent</u> if and only if $\sum_{i=1}^{r} n_i g_i = 0$ for $n_i \in \mathbf{Z}$ implies that $n_i = 0$ for all i. The supremum of the r for which there is an independent set of r elements is called the <u>rank</u> of G. Show that the rank of $\mathbf{Q}^*$ is ∞.

11. Now let K/k be a field extension, and $|\ |$ be a valuation on K. Let $\bar{K}/\bar{k}$ be the corresponding residue class field extension; show that $\mathfrak{p}$

$$[K:k] \geq \operatorname{rank}(|K^*|/|k^*|) + [\bar{K}:\bar{k}].$$

[Hint: in 9 take K to be the valuation ring of $|\ |$, and to be its maximal ideal.] [Further hint: use 2.]

12. If K/k is an algebraic extension (i. e. $[K:k] = 0$ so that $\bar{K}/\bar{k}$ is an algebraic extension), there is a curious analogue of 11 (though right-minded people would say that 11 was a curious analogue of 12). Define $|K:k|$ to be the dimension of K as a k-vector space (perhaps infinite), and show:

$$|K:k| \geq |\bar{K}:\bar{k}| \, \Big| \, |K^*| : |k^*| \, \Big| \, .$$

More examples can be found in Bourbaki, <u>Alg. Comm.</u>, V.

8. Prüfer and Dedekind rings

Throughout this chapter A denotes an integral ring and k its field of fractions.

8.1 (Lemma). *Let A be a valuation ring and M be a torsion-free A-module of finite type. Then M is free.*

Proof. Let $(m_i)_{1 \le i \le n}$ generate M for some *least* $n \in \omega$. Suppose $\sum_{i=1}^{n} \xi_i m_i = 0$ for $\xi_i \in A$ with (say) $\xi_n \in A^*$ and $|\xi_n| \ge |\xi_i|$ for $i = 1, \ldots, n$. Then $\xi_n^{-1}\xi_i \in A$ for all i and $\xi_n(m_n + \sum_{i=1}^{n-1} (\xi_n^{-1}\xi_i)m_i) = 0$: so that $m_n = -\sum_{i=1}^{n-1} (\xi_n^{-1}\xi_i)m_i$ and $(m_i)_{1 \le i \le n-1}$ generates M, a contradiction. Thus $\xi_i = 0$ for all i and $(m_i)_{1 \le i \le n}$ bases M. □

8.2 (Theorem). *The following four conditions are equivalent:*

(i) *$A_{\mathfrak{p}}$ is a valuation ring for every $\mathfrak{p} \in \operatorname{spec}(A)$;*

(ii) *every torsion-free A-module of finite type is projective;*

(iii) *if $\mathfrak{a}$ is a non-zero ideal of A of finite type, then $\mathfrak{a}$ is an invertible sub-A-module of k;*

(iv) *every sub-A-algebra of k is integrally closed in k.*

Proof. Suppose (i) holds. Let M be a torsion-free A-module of finite type. Then $M_{\mathfrak{p}}$ is a torsion-free $A_{\mathfrak{p}}$-module of finite type for every $\mathfrak{p} \in \text{spec}(A)$, so that by 8.1 $M_{\mathfrak{p}}$ is a free $A_{\mathfrak{p}}$-module, of rank $r_{\mathfrak{p}}$ say. However $M_{\{0\}}$ is isomorphic to $M_{\mathfrak{p}} \otimes_{A_{\mathfrak{p}}} k$ for all $\mathfrak{p} \in \text{spec}(A)$, so by 1.3.1 $r_{\mathfrak{p}} = r_{\{0\}}$ for all $\mathfrak{p} \in \text{spec}(A)$, and M is projective by 3.3.7 (iii). That is, (ii) holds.

Suppose (ii) holds and let us prove (iii). By (ii) $\mathfrak{a}$ is projective, and thus invertible by 3.4.2 and 3.4.3.

Suppose (iii) holds and let $\mathfrak{p} \in \text{spec}(A)$ and $x \in k \backslash A_{\mathfrak{p}}$. Then $x = \frac{y}{z}$ for $y, z \in A$ with $z \neq 0$ and $Ay + Az$ is invertible; so $A + Ax$ is invertible (for its inverse is $z(A:Ay + Az)$). Thus $1 \in (A + Ax)(A:A + Ax)$ so that $1 = a + xb$ for some $a, b \in A$ with $ax, bx \in A$. Suppose $a \in \mathfrak{u}(A_{\mathfrak{p}})$: then $x = a^{-1}ax \in A_{\mathfrak{p}}$, a contradiction. Thus $a \in \mathfrak{m}(A_{\mathfrak{p}})$ and $x^{-1} = (1 - a)^{-1} b \in A_{\mathfrak{p}}$. That is, (i) holds.

Suppose (i) holds and let B be a sub-A-algebra of k and $\mathfrak{m}$ be a maximal ideal of B. Then $\mathfrak{m} \cap A = \mathfrak{p} \in \text{spec}(A)$ and $A_{\mathfrak{p}} \leq B_{\mathfrak{m}}$. Thus by 7.3.1 $A_{\mathfrak{p}} = B_{\mathfrak{m}}$ and $B_{\mathfrak{m}}$ is a valuation ring in k and is therefore integrally closed in k. But by 3.2.1.3 $B = \bigcap_{\mathfrak{m}} B_{\mathfrak{m}}$ and thus B is integrally closed in k: that is, (iv) holds.

Finally suppose (iv) holds; let $\mathfrak{p} \in \text{spec}(A)$; and $x \in k \backslash A_{\mathfrak{p}}$. Then $A_{\mathfrak{p}}[x^2]$ is integrally closed in k, so $x \in A_{\mathfrak{p}}[x^2]$. Thus $x = b_0 + b_1x^2 + \ldots + b_nx^{2n}$ for some $b_0, \ldots, b_n \in A_{\mathfrak{p}}$ with $n > 0$; and

$$(b_0x^{-1})^{2n} - (b_0x^{-1})^{2n-1} + b_0^2b_1(b_0x^{-1})^{2n-2} + \ldots + b_0^{2n}b_n = 0$$

so that $b_0 x^{-1} \in A_{\mathfrak{p}}$. However $b_0 x^{-1} \in \mathfrak{m}(A_{\mathfrak{p}})$ for if not, $x = b_0(b_0 x^{-1})^{-1} \in A_{\mathfrak{p}}$; and so

$$(x^{-1})^{2n-1} + (b_0 x^{-1}-1)^{-1} b_1 (x^{-1})^{2n-2} + \ldots + (b_0 x^{-1}-1)^{-1} b_n = 0$$

and $x^{-1} \in A_{\mathfrak{p}}$. Thus (i) holds. □

If A satisfies the conditions of 8.2 we call A a Prüfer ring. For example $\mathbf{Z}$ is a Prüfer ring.

8.2.1a (Corollary). We have a further equivalent condition:

(v) $A + Ax$ is invertible for all $x \in k$.

Proof. Both (iii) implies (v) and (v) implies (i) are already contained in the proof that (iii) implies (i). □

8.2.1 (Corollary). If A is a Prüfer ring, every sub-A-algebra B of k is a Prüfer ring.

Proof. Let $\mathfrak{P} \in \operatorname{spec}(B)$. Then $A_{\mathfrak{P} \cap A} \leq B_{\mathfrak{P}}$ and $B_{\mathfrak{P}}$ is a valuation ring in k by 7.3.1. □

8.2.2 (Corollary). Let A be a Prüfer ring and K be an algebraic extension of k. Let B be the integral closure of A in K. Then B is a Prüfer ring.

Proof. By 5.1.3 and 7.3.4. □

Exercises. Find (i) the rings A for which 8.1 is true, (ii) the subrings of **Q**, (iii) a Prüfer ring which is not Bezout.

We say that a ring A is a Dedekind ring iff it is a Noetherian Prüfer ring. For example **Z** is a Dedekind ring.

8.3 **(Theorem).** The following four conditions are equivalent:

(i) A is a Dedekind ring;

(ii) A is Noetherian and integrally closed in k; and if $\mathfrak{p} \in \operatorname{spec}(A)$, then $\mathfrak{p} = 0$ or $\mathfrak{p}$ is maximal;

(iii) A is Noetherian and $A_{\mathfrak{m}}$ is a field or a discrete valuation ring for every maximal ideal $\mathfrak{m}$ of A;

(iv) if $\mathfrak{a}$ is a non-zero ideal of A, then $\mathfrak{a}$ is an invertible sub-A-module of k.

Proof. Suppose (i) holds. Then A is Noetherian and integrally closed by 8.2 (iv). Suppose $\mathfrak{p} \in \operatorname{spec}(A)$ and $0 \subset \mathfrak{p} \subset \mathfrak{m}$ for some maximal ideal $\mathfrak{m}$ of A. Then $A_{\mathfrak{m}} \subset k$ and $\operatorname{card}(\operatorname{spec}(A_{\mathfrak{m}})) \geq 3$ and $A_{\mathfrak{m}}$ is a Noetherian valuation ring: this is a contradiction by 7.2.3 (iii), (iv). Thus (ii) holds.

Suppose (ii) holds. Then for any non-zero maximal ideal $\mathfrak{m}$ of A, we have $A_{\mathfrak{m}} \subset k$ and $A_{\mathfrak{m}}$ is a Noetherian ring, integrally closed in k by 5.1.3, and with $\operatorname{card}(\operatorname{spec}(A_{\mathfrak{m}})) = 2$. Thus by 7.2.3 (iv) $A_{\mathfrak{m}}$ is a discrete valuation ring in k: that is, (iii) holds.

Suppose (iii) holds. Let $\mathfrak{p} \in \operatorname{spec}(A)$ so that $\mathfrak{p} \subseteq \mathfrak{m}$ for some maximal $\mathfrak{m}$. Then $A_{\mathfrak{m}} \subseteq A_{\mathfrak{p}}$ so that $A_{\mathfrak{p}}$ is a valuation ring, and A is a Prüfer ring by 8.2 (i). Thus (i) holds.

Suppose (i) holds. Then (iv) holds by 8.2 (iii).

Finally suppose (iv) holds. Then A is a Prüfer ring by 8.2 (iii); and since an invertible sub-A-module of k is of finite type, A is Noetherian. Thus (i) holds. □

Suppose $A \subset k$ is a Dedekind ring. Then $A_{\mathfrak{m}}$ is a discrete valuation ring in k for each maximal ideal $\mathfrak{m}$ of A, and we write $\mathrm{ord}_{\mathfrak{m}} : k^* \to \mathbf{Z}$ for the corresponding map.

Let $\mathfrak{a}$ be an invertible sub-A-module of k. Then $\mathfrak{a}$ is of finite type, so that $x\,\mathfrak{a} \subseteq A$ for some $x \in A^*$, and we can define

$$\mathrm{ord}_{\mathfrak{m}}(\mathfrak{a}) = \inf_{0 \neq x \in \mathfrak{a}} \mathrm{ord}_{\mathfrak{m}}(x) .$$

Thus if G is the group of invertible sub-A-modules of k, then $\mathrm{ord}_{\mathfrak{m}} : G \to \mathbf{Z}$ is a group morphism; and so we have a group morphism: $G \to \prod_{\mathfrak{m}} \mathbf{Z}$ given by $\mathfrak{a} \longmapsto (\mathrm{ord}_{\mathfrak{m}}(\mathfrak{a}))_{\mathfrak{m}}$. In particular if $\mathfrak{n}$ is a maximal ideal we have:

$$\begin{aligned} \mathrm{ord}_{\mathfrak{m}}(\mathfrak{n}) &= 0 \quad \text{if } \mathfrak{m} \neq \mathfrak{n} \\ &= 1 \quad \text{if } \mathfrak{m} = \mathfrak{n} . \end{aligned}$$

8.4 (Theorem). *The map:* $G \to \prod_{\mathfrak{m}} \mathbf{Z}$ *just defined is a group isomorphism:* $G \to \bigoplus_{\mathfrak{m}} \mathbf{Z}$.

Proof. Let $x \in A^*$ and $(\mathfrak{m}_i)_{i \in \omega}$ be a sequence of *distinct* maximal ideals of A with $x \in \mathfrak{m}_i$ for all i. Then $A \supset \mathfrak{m}_1 \supset \mathfrak{m}_1 \cap \mathfrak{m}_2 \supset \ldots \supseteq xA$, for if $\mathfrak{m}_1 \cap \ldots \cap \mathfrak{m}_r = \mathfrak{m}_1 \cap \ldots \cap \mathfrak{m}_{r-1}$ we have $\mathfrak{m}_r \supseteq \mathfrak{m}_1 \ldots \mathfrak{m}_{r-1}$ and $\mathfrak{m}_r = \mathfrak{m}_i$ for some $i < r$. Thus A/xA is *not* Artinian and *not* of finite length; but A/xA is a torsion A-module of finite type: a contradiction by 8. 3 (ii) and 4. 3. 5 (i). Thus $x \notin \mathfrak{m}$ and $\mathrm{ord}_{\mathfrak{m}}(x) = 0$ for all but a finite number of $\mathfrak{m}$.

Clearly this also holds for $x \in k^*$.

Let $\mathfrak{a} \in G$. Then $\mathfrak{a} = Ax_1 + \dots + Ax_n$ for some $x_1, \dots, x_n \in k^*$, so that:

$$\mathrm{ord}_{\mathfrak{m}}(\mathfrak{a}) = \inf_{i=1}^{n} \mathrm{ord}_{\mathfrak{m}}(x_i) = 0$$

for all but a finite number of $\mathfrak{m}$.

Thus $G \to \bigoplus_{\mathfrak{m}} \mathbf{Z}$.

Moreover $\prod_{\mathfrak{m}} \mathfrak{m}^{r_{\mathfrak{m}}} \mapsto (r_{\mathfrak{m}})_{\mathfrak{m}}$ for any family $(r_{\mathfrak{m}})_{\mathfrak{m}}$ of finite support in $\mathbf{Z}$: so that $G \xrightarrow{\text{onto}} \bigoplus_{\mathfrak{m}} \mathbf{Z}$.

Finally let $\mathfrak{a} \in G$ and $\mathrm{ord}_{\mathfrak{m}}(\mathfrak{a}) = 0$ for all $\mathfrak{m}$. Then $\mathfrak{a}_{\mathfrak{m}} = A_{\mathfrak{m}}$ for each $\mathfrak{m}$, for $A_{\mathfrak{m}}$ is a discrete valuation ring; so that $\mathfrak{a}_{\mathfrak{m}} = \mathfrak{a}_{\mathfrak{m}} + A_{\mathfrak{m}} = (\mathfrak{a} + A)_{\mathfrak{m}} = A_{\mathfrak{m}}$ for all $\mathfrak{m}$; and by 3. 2. 1. 1 applied to the inclusions: $\mathfrak{a}, A \to \mathfrak{a} + A$ we have $\mathfrak{a} = A + \mathfrak{a} = A$. That is, $G \xrightarrow{\text{inj}} \bigoplus_{\mathfrak{m}} \mathbf{Z}$. □

Thus every non-zero ideal of a Dedekind ring is uniquely expressible as a product of maximal ideals. The converse is true too:

8. 4. 1 (Corollary). $A \subset k$ is a Dedekind ring iff every non-zero ideal of A is uniquely expressible as a product of prime ideals.

Proof. One way is 8. 4.

Conversely let $\mathfrak{p} \in \operatorname{spec}(A)$ and $\mathfrak{p} \neq 0$. Define $\operatorname{ord}_{\mathfrak{p}}(x)$ for $x \in A^*$ to be the exponent of $\mathfrak{p}$ in the expression of xA as a product of prime ideals. Clearly

$$\operatorname{ord}_{\mathfrak{p}}(xy) = \operatorname{ord}_{\mathfrak{p}}(x) + \operatorname{ord}_{\mathfrak{p}}(y)$$

$$\operatorname{ord}_{\mathfrak{p}}(x + y) \geq \inf(\operatorname{ord}_{\mathfrak{p}}(x), \operatorname{ord}_{\mathfrak{p}}(y))$$

$$x \in A \setminus \mathfrak{p} \text{ iff } \operatorname{ord}_{\mathfrak{p}}(x) = 0$$

for all $x, y \in A$. We define $\operatorname{ord}_{\mathfrak{p}} : k^* \to \mathbf{Z}$ by $\operatorname{ord}_{\mathfrak{p}}(\frac{x}{y}) = \operatorname{ord}_{\mathfrak{p}}(x) - \operatorname{ord}_{\mathfrak{p}}(y)$ for $x, y \in A^*$. Then we see that $A_{\mathfrak{p}}$ is a discrete valuation ring and that $\operatorname{ord}_{\mathfrak{p}} : k^* \to \mathbf{Z}$ is the associated map.

By 8.2 (i) it remains to show that A is Noetherian. Since

$$\prod_{\mathfrak{p} \neq 0} \mathfrak{p}^{m_{\mathfrak{p}}} \supseteq \prod_{\mathfrak{p} \neq 0} \mathfrak{p}^{n_{\mathfrak{p}}} \text{ iff } m_{\mathfrak{p}} \leq n_{\mathfrak{p}} \text{ for all } \mathfrak{p} \neq 0,$$

there are exactly $\prod_{\mathfrak{p} \neq 0} (n_{\mathfrak{p}} + 1)$ ideals between A and $\prod_{\mathfrak{p} \neq 0} \mathfrak{p}^{n_{\mathfrak{p}}}$; so A satisfies the increasing sequence condition and is Noetherian. □

8.5 (Theorem) (Krull-Akizuki). Let $A \subset k$ be a Dedekind ring; K be a finite field extension of k; and B be the integral closure of A in K. Then B is a Dedekind ring.

Proof. From 8.2.2 B is a Prüfer ring: thus it remains to show that B is Noetherian. For this it certainly suffices to show that $B/\mathfrak{b}$ is an A-module of finite length for any non-zero ideal $\mathfrak{b}$ of B.

Let $y \in \mathfrak{b}$ and $y \neq 0$. Then $y^n + a_1 y^{n-1} + \ldots + a_n = 0$ for some $a_1, \ldots, a_n \in A$ and least $n \in \omega$, so that $a_n \in A^*$. Thus

$a_n B \subsetneq \mathfrak{b}$ and it is enough to show that $B/a_n B$ is an A-module of finite length.

We have $B \otimes_A k \xrightarrow{\text{iso}} K$ by $x \otimes y \longmapsto xy$; so B is a torsion-free A-module of finite rank $|K:k|$. Thus by 4. 3. 5 (ii) $B/a_n B$ is an A-module of finite length. □

Exercises. (i) A Dedekind ring with only finitely many maximal ideals is local.

(ii) If $\mathfrak{a}$ is an ideal of a Dedekind ring A, there exist $x, y \in A$ such that $\mathfrak{a} = Ax + Ay$.

9. General exercises

1. Recall that in any category, a morphism $f:A \to B$ is called an epimorphism iff given two morphisms $g, h:B \to C$ for some C with $gf = hf$, it follows that $g = h$; in other words a morphism with domain B is determined by its composition with f.

Show that for $f:A \to B$ a ring morphism the following six conditions are equivalent:

(i) f is an epimorphism

(ii) $B \otimes_A B \xrightarrow{\text{inj}} B$ by $b \otimes b' \longmapsto bb'$

(iii) $B \xrightarrow{\text{onto}} B \otimes_A B$ by $b \longmapsto b \otimes 1$

(iv) $B \xrightarrow{\text{onto}} B \otimes_A B$ by $b \longmapsto 1 \otimes b$

(v) for all $b \in B$ $b \otimes 1 = 1 \otimes b$ in $B \otimes_A B$

(vi) $B \otimes_A B \xrightarrow{\text{iso}} B$ by multiplication.

2. Let $A \to B$ be a ring morphism. Show the equivalence of:

(i) $A \xrightarrow{\text{epi}} B$

(ii) if M, N are B-modules and $f:M \to N$ is an A-module morphism, then it is also a B-module morphism

(iii) if M, N are B-modules then $M \otimes_A N \xrightarrow{\text{iso}} M \otimes_B N$ by $m \otimes n \longmapsto m \otimes n$.

3. (i) Let A be a ring and S a multiplicative subset of A. Show that $A \xrightarrow{\text{epi}} S^{-1}A$.

(ii) Let A be a ring and $a \in A$. Show that $A \xrightarrow{\text{epi}} A/aA \times A_a$ by $x \longmapsto (x + aA, \frac{x}{1})$.

[Hint: use 1 (vi).] Use this to construct an epimorphism not of the form (i) above. [Hint: take $A = \mathbf{Z}$, $a \neq 0, 1$.]

4. Let $f: A \to B$ be a flat epimorphism of rings (that is $A \xrightarrow{\text{epi}} B$ and B is a flat A-module). Show that $\operatorname{spec}(B) \xrightarrow{\text{embed}} \operatorname{spec}(A)$; indeed, for any ideal $\mathfrak{b}$ of B $f^{-1}[\mathfrak{b}]B = \mathfrak{b}$.

5. Give a proof or counter-example for each of the following statements:

(i) let A be an integral ring and make $A \otimes_{\mathbf{Z}} A$ into an A-module by $\lambda(x \otimes y) = (\lambda x) \otimes y$ for $\lambda, x, y \in A$; then $A \otimes_{\mathbf{Z}} A$ is faithfully flat;

(ii) let A be a Prüfer ring and B be the integral closure of A in an algebraic extension K of the field of fractions k of A; then B is a faithfully flat A-module.

6. Which of the following properties of a ring A remain true for $S^{-1}A$ for any multiplicative subset S of A? Give counter-examples or proofs.

(i) A is Noetherian;

(ii) $\operatorname{spec}(A)$ is connected;

(iii) every A-module is flat;

(iv) A is a valuation ring or a zero ring;

(v) A has zero nilradical;

(vi) A is of finite type as a $\mathbf{Z}$-algebra.

7. Let A be an integral ring of finite type as a $\mathbf{Z}$-algebra. Prove, or give a counter-example for each of the following statements:

(ii) if A is a valuation ring, A is a field;

(iii) if A is a local ring, A is a valuation ring.

What happens if we allow A to have divisors of zero?

8. Give proofs or counter-examples for the following statements:

(i) if B is an entire A-algebra, and every A-module is flat, and B has no nilpotent elements, then every B-module is flat;

(ii) if A is integral and integrally closed in its field of fractions, then A has unique factorisation;

(iii) let B be a ring integrally dependent on a local subring; then B is local.

Appendix 1

A NOTE ON CATEGORIES, FUNCTORS AND NATURAL TRANSFORMATIONS

(The definitions here need a little set-theoretic polishing, but this would make them much longer: see the introduction to Freyd's Abelian Categories.)

A category $\mathscr{A}$ consists of

(i) a class of objects A, B, C, . . .

(ii) for every two objects A, B a set $[A, B]_{\mathscr{A}}$ of morphisms $\alpha, \beta, \ldots$ from A to B (and we write $\alpha: A \to B$ to mean $\alpha \in [A, B]_{\mathscr{A}}$)

(iii) for every three objects A, B, C a multiplication taking $[B, C]_{\mathscr{A}} \times [A, B]_{\mathscr{A}}$ to $[A, C]_{\mathscr{A}}$;

and must also satisfy the following conditions:

(i) (associativity) if $\alpha: A \to B$ and $\beta: B \to C$ and $\gamma: C \to D$, then $\gamma(\beta\alpha) = (\gamma\beta)\alpha$

(ii) (existence of identities) for each object A there is a morphism $e_A: A \to A$ such that if $\alpha: A \to B$ (resp. $\beta: C \to A$) we have $\alpha e_A = \alpha$ (resp. $e_A\beta = \beta$)

(iii) if $[A, B]_{\mathscr{A}}$ meets $[C, D]_{\mathscr{A}}$, then $A = C$ and $B = D$.

EXAMPLES

1. The category Ring of (commutative) rings (with ones):

(i) the objects are rings;

(ii) if A and B are rings, $[A, B]_{Ring}$ consists of the ordered triples (A, B, f) where f is a function from A to B satisfying

$$f(1) = 1$$

and

$$f(x + y) = f(x) + f(y)$$
$$f(xy) = f(x)f(y)$$

for all $x, y \in A$;

(iii) if $(A, B, f): A \to B$ and $(B, C, g): B \to C$, then $(B, C, g)(A, B, f) = (A, C, g \circ f): A \to C$.

2. The category Top of topological spaces:

(i) the objects are topological spaces;

(ii) if S and T are topological spaces, then $[S, T]_{Top}$ consists of the ordered triples (S, T, f) where f is a continuous function from S to T;

(iii) the multiplication is just as for rings.

3. The dual category $\mathscr{A}^*$ of any category $\mathscr{A}$:

(i) the objects of $\mathscr{A}^*$ are those of $\mathscr{A}$;

(ii) $[A, B]_{\mathscr{A}^*} = [B, A]_{\mathscr{A}}$;

(iii) $[B, C]_{\mathscr{A}^*} \times [A, B]_{\mathscr{A}^*} \to [A, C]_{\mathscr{A}^*}$ is $(\beta, \alpha) \longmapsto \alpha\beta$.

(Thus $(\mathscr{A}^*)^* = \mathscr{A}$.)

Now let $\mathscr{A}$ and $\mathscr{B}$ be categories and F be a function from the objects of $\mathscr{A}$ to those of $\mathscr{B}$ and from the morphisms of $\mathscr{A}$ to those of $\mathscr{B}$. We say that F is a functor from $\mathscr{A}$ to $\mathscr{B}$ iff:

(i) $F[[A, B]_{\mathscr{A}}] \subseteq [F(A), F(B)]_{\mathscr{B}}$;

(ii) $F(e_A) = e_{F(A)}$ for every object A of $\mathscr{A}$;

(iii) if $\alpha:A \to B$ and $\beta:B \to C$ in $\mathscr{A}$, then $F(\beta\alpha) = F(\beta)F(\alpha)$.

Strictly speaking this is a covariant functor; a contravariant functor from $\mathscr{A}$ to $\mathscr{B}$ is just a covariant functor from $\mathscr{A}$ to $\mathscr{B}^*$. Thus F is a contravariant functor from $\mathscr{A}$ to $\mathscr{B}$ iff:

(i)* $F[[A, B]_{\mathscr{A}}] \subseteq [F(B), F(A)]_{\mathscr{B}}$;

(ii)* same as (ii) above;

(iii)* if $\alpha:A \to B$ and $\beta:B \to C$ in $\mathscr{A}$, then $F(\beta\alpha) = F(\alpha)F(\beta)$.

EXAMPLES

1. The identity function on a category $\mathscr{A}$.

2. The contravariant functor spec from Ring to Top; if $(A, B, f):A \to B$ where A and B are rings, we define

$$\text{spec}(A, B, f) = (\text{spec}(B), \text{spec}(A), g):\text{spec}(B) \to \text{spec}(A)$$

where g is the (continuous) function from spec(B) to spec(A) given by

$$g(\mathfrak{p}) = f^{-1}[\mathfrak{p}]$$
$$= \{x \in A : f(x) \in \mathfrak{p}\}$$

for $\mathfrak{p} \in \mathrm{spec}(B)$; it is easy to check that

$$\mathrm{spec}(\beta\alpha) = \mathrm{spec}(\alpha)\mathrm{spec}(\beta)$$

for $\alpha : A \to B$ and $\beta : B \to C$.

Now let F and G be functors from $\mathscr{A}$ to $\mathscr{B}$. A *natural transformation* χ from F to G is a function taking *objects* of $\mathscr{A}$ to *morphisms* of $\mathscr{B}$ such that

(i) $\chi(A) : F(A) \to G(A)$ for all objects A of $\mathscr{A}$;

(ii) if $\alpha : A \to B$ in $\mathscr{A}$, then the diagram in $\mathscr{B}$

$$\begin{array}{ccc} F(A) & \xrightarrow{\chi(A)} & G(A) \\ \Big\downarrow{\scriptstyle F(\alpha)} & & \Big\downarrow{\scriptstyle G(\alpha)} \\ F(B) & \xrightarrow{\chi(B)} & G(B) \end{array}$$

commutes (i. e. $G(\alpha)\chi(A) = \chi(B)F(\alpha)$).

EXAMPLES

1. If $F = G$ we can define $\chi(A) = e_{F(A)}$.

2. (The *double dual.*) Let Mod be the category of modules over a fixed ring A. We have a contravariant functor F

from Mod to Mod: we define $F(M)$ for each module M to be $[M, A]_{Mod}$ made into an A-module in the obvious way, and if $(M, N, f): M \to N$ in Mod, we have $F(M, N, f) = (F(N), F(M), g): F(N) \to F(M)$ where g is a function from $F(N)$ to $F(M)$ given by $g(\alpha) = \alpha(M, N, f)$ for all $\alpha: N \to A$. Then we have a natural transformation χ from 1_{Mod} to $F \circ F$ (which is a (covariant) functor): namely

$$\chi(M) = (M, F(F(M)), h): M \to F(F(M)) ,$$

where $h(m) = (F(M), A, j): F(M) \to A$, where $j(M, A, f) = f(m)$ for all $m \in M$ and $(M, A, f): M \to A$.

Appendix 2

THE CONSTRUCTIBLE TOPOLOGY

(We describe topologies by their closed sets.)

Let A be a ring and σ be the least topology on spec(A) such that D(f) is both open and closed for all $f \in A$. Then σ is Hausdorff and contains the Zariski topology.

If B is an A-algebra write $\zeta(B)$ for the image of spec(B) in spec(A) and let

$$\tau = \{\zeta(B) : B \text{ an A-algebra}\}.$$

Recall that if $(B_i)_{i \in I}$ is a direct family of A-algebras with maps $f_{ij}: B_i \to B_j$ for $i \leq j$ we have the direct limit $B = \varinjlim_{i \in I} B_i$ defined as disjoint $\bigcup_{i \in I} B_i/\sim$ where $(x, i) \sim (y, j)$ iff $f_{il}(x) = f_{jl}(y)$ for some $l \geq i, j$. We make B into an A-algebra in an obvious way and let $B_i \to B$ by $x \mapsto$ class of (x, i). If C is another A-algebra we have:

$$C \otimes_A \varinjlim_{i \in I} B_i \xrightarrow[\text{iso}]{\text{nat}} \varinjlim_{i \in I} C \otimes_A B_i .$$

Theorem. (i) $\zeta(B \otimes_A C) = \zeta(B) \cap \zeta(C)$

(ii) $\zeta(B \times C) = \zeta(B) \cup \zeta(C)$

(iii) $\zeta(\varinjlim_{i \in I} B_i) = \bigcap_{i \in I} \zeta(B_i)$

(iv) $\zeta(\bigotimes^{res}_{\lambda \in \Lambda} B_\lambda) = \bigcap_{\lambda \in \Lambda} \zeta(B_\lambda)$

(v) $\tau = \sigma$ and is a compact Hausdorff topology on spec(A).

Proof. Looking at

$$\begin{array}{ccc} A & \longrightarrow & B \\ \downarrow & & \downarrow \\ k & \longrightarrow & k \otimes_A B \end{array}$$

where $k = A_{\mathfrak{p}} / \mathfrak{p} A_{\mathfrak{p}}$ for some $\mathfrak{p} \in$ spec(A) we see that $\mathfrak{p} \in \zeta(B)$ iff $k \otimes_A B \neq 0$. But $(k \otimes_A B) \otimes_A (k \otimes_A C)$ is isomorphic to $k \otimes_A (B \otimes_A C)$ since $A \xrightarrow{\text{epi}} k$; and this gives (i). Similarly we get (ii), (iii) and (iv) from the natural isomorphisms:

$$(B \times C) \otimes_A k \to (B \otimes_A k) \times (C \otimes_A k)$$

$$(\varinjlim_{i \in I} B_i) \otimes_A k \to \varinjlim_{i \in I} (B_i \otimes_A k)$$

$$\bigotimes^{res}_{\lambda \in \Lambda} B_\lambda \to \varinjlim_{\substack{\Lambda_0 \subseteq \Lambda \\ \Lambda_0 \text{ finite}}} (\bigotimes_{\lambda \in \Lambda_0} B_\lambda) .$$

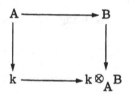

Thus τ is a topology. If $\bigcap_{\lambda \in \Lambda} \zeta(B_\lambda) = \emptyset$, then $k \otimes_A (\bigotimes_{\lambda \in \Lambda_0} B_\lambda) = 0$ for some finite $\Lambda_0 \subseteq \Lambda$, so $\bigcap_{\lambda \in \Lambda_0} \zeta(B_\lambda) = \emptyset$ and τ is compact. We have $D(f) = \zeta(A_f) \in \tau$ and $X \backslash D(f) = \zeta(A/fA) \in \tau$ so that $\tau \supseteq \sigma$ and by analysis $\tau = \sigma$. □

We call τ the constructible topology on spec(A).

Bibliography

For reference:

Bourbaki	Algèbre Commutative (Hermann)
Grothendieck	Éléments de Géométrie Algébrique (IHES)
Zariski and Samuel	Commutative Algebra (Van Nostrand 1958 and 1960)

For reading and exercises:

Atiyah and Macdonald	Introduction to Commutative Algebra (Addison-Wesley 1969)
Kaplansky	Commutative Rings (QMC Notes)

For specialised topics:

Serre	Corps Locaux (Hermann 1968)
Artin	Theory of Algebraic Numbers (Göttingen 1959)
Samuel (ed.)	Les épimorphismes d'anneaux (Paris 1968)
Cohen and Seidenberg	Bull. Amer. Math. Soc. 52 (1946), 252-61
Nagata	Local Rings (Interscience 1962)

Index of notation

Index of terms

Printed in the United States
By Bookmasters